92790

# CREATIVE CAREERS
# TV

**IDEAS**FACTORY

## Milly Jenkins

trotman

*Creative Careers: TV*
This first edition published in 2003 by Trotman and Company Ltd
2 The Green, Richmond, Surrey TW9 1PL

© Trotman and Company Limited 2003

Produced in association with Channel 4 **IDEAS**FACTORY

**Editorial and Publishing Team**
**Author** Milly Jenkins
**Editorial** Mina Patria, Editorial Director; Rachel Lockhart, Commissioning
Editor; Anya Wilson, Editor; Erin Milliken, Editorial Assistant
**Production** Ken Ruskin, Head of Pre-press and Production
**Sales and Marketing** Deborah Jones, Head of Sales and Marketing
**Managing Director** Toby Trotman

**Designed by** XAB

British Library Cataloguing in Publication Data
A catalogue record for this book is available from the British Library

ISBN  0 85660 901 3

Typeset by Mac Style Ltd, Scarborough, N. Yorkshire
Printed and bound in Great Britain by The Cromwell Press,
Trowbridge, Wiltshire

# Contents

## ABOUT THE AUTHOR

Milly Jenkins studied history at Manchester University and then broadcast journalism at Cardiff University. She started her career as a researcher for BBC Radio Wales, going on to work on news and current affairs documentaries for BBC Radio 4 and 5. She has also worked as a writer and editor at *The Guardian* and is now a freelance journalist, specialising in careers and working life. She has written for the *Independent*, *The Times*, the *Financial Times*, the *Evening Standard* and *Marie Claire*. She lives in London.

## ACKNOWLEDGEMENTS

Thanks to Harriet Messenger for additional research. Also, to Graham Stuart and Nina Gosling at So Television, Ian Lewis at the Farnham Film Company, and to Sophie Hutchinson, Faye Hawker, Jane Bussmann and David Quantick, to Adam Gee, James Estill, Julian Mobbs and Katie Streten at Channel 4's **IDEAS**FACTORY, Akeva K Avery at Keep-up-to-date productions, Jonnie Turpie from Maverick TV, Mundy Ellis – editor of *Televisual*, Tammy Learn of 4Active and to Mina Patria, Anya Wilson and Rachel Lockhart at Trotman.

Most of all, a big thank you to all the people who work in TV who agreed to be interviewed for this book.

'You will meet hundreds of people, go to a hundred different places and have the immense satisfaction of seeing the programmes you have made on screen.'

## what this book will do for you

You're not alone. It is estimated that 60,000 new people try to break into the UK's audio-visual industries – TV, Radio, Film and Interactive Media – every year. That's a lot, considering 200,000 people already work in them.[1]

Of those 60,000 young hopefuls, it's safe to assume that the majority of them want to work in TV – *the* most popular destination in the media. Is there enough work for them out there?

No. According to Skillset, an industry skills council that does an annual employment census, there were about 60,000 jobs in TV in 2002, with broadcasters and independent production companies, as well as in other TV-related sectors – commercials, corporate, 'non-broadcast' production, facilities, animation and special effects.

So are all those wanting to work in TV wasting their time? Of course not. Although TV is a tough business to get into, and a tough business to stay in, there is always a demand for new talent and skilled professionals. This is an industry that eats up ideas (and people).

The trouble is, although the industry always needs fresh blood, and there are thousands queuing up to give it, there aren't actually that many people who have what it takes. Employers say too many people they interview have starry-eyed notions about the 'glamour' of working in TV. They don't understand how the industry works and what different jobs involve. Some don't even really watch TV.

So you need to know as much about the industry as possible, before you decide whether or not it's for you. TV is a famously insecure profession – there is very little stability. But, if you go in with your eyes wide open, understanding the business and how tough it can be, you'll have a good time. You will meet hundreds of people, go to a hundred different places and have the immense satisfaction of seeing the programmes you have made on screen.

## WHAT THIS BOOK WILL DO FOR YOU

This book is a guide to working in TV. It will tell you about the state of the TV industry, where it has got to and where it is going. It will tell you how

# Insiders tell it like it really is

'I've done it for 25 years and I can't imagine doing anything else. It's a great place for creative team players. You get to meet like-minded people and make something together that can touch people's lives. It's a fantastic job.'

*Graham Stuart, Executive Producer of*
V Graham Norton *and Director of* So Television

'The best thing about TV is the amount of people your work reaches. Sometimes it's the people you get to work with – with the right team it's exhilarating. The worst thing is getting the gig in the first place. Most people I know spend at least half their time in development hell.'

*Meera Syal, Actress, Novelist and Screenwriter*

'The average day doesn't exist – which is part of the fun! Last week I was filming Linford Christie one day and camping in a field the next.'

*Louise Doolan, Assistant Producer*

the industry works, who the big employers are and what working in TV is like on a day-to-day basis: what the working environment is like, and what the people and the money are like.

It will also explain who does what in TV, helping you to think about the kind of jobs you might be interested in and suited to, as well as telling you how to train for them and find work in them. You'll hear from people already in those jobs, talking about what they love and hate about the industry and what they love and hate about their work.

Like the rest of the Creative Careers series on working in the media – in television, film and radio – the aim is to give you the real deal on what it's like to work in TV, highlighting the positives and negatives in what can be a challenging but rewarding business.

## NOTE

1. *Skills for Tomorrow's Media: The Report of the Skillset/DCMS Audio Visual Industries Training Group*, September 2001.

'I enjoy every minute of it. Everything about television is the best, but the very best is meeting people I've admired my whole life. The worst is learning my lines. I hate it!'

*Vernon Kay, Presenter*

'The real satisfaction and aim is to make something that is going to entertain the viewer. The downside is the long hours, endless discussions about budgets… and the fact that sometimes, despite your best efforts, what you make doesn't hit the mark.'

*Francis Hopkinson, Drama Producer*, The Cops, A&E, The Jury

'It's very insecure. In Britain, unlike the US, there's no set career route, with long-running sitcoms that writers can train on. The upside is that there is always the possibility, albeit slim, of getting your own show on.'

*Jesse Armstrong, Comedy Writer, including*
Peep Show *and* Smack the Pony

'Britain is a TV nation. Watching TV is the most popular leisure activity.'

# a (very) brief history of british television

## THE PAST

A (very) brief history of British television.

*1930s:* The BBC makes its first experimental broadcast in 1932 – with a 30-mile transmitter range. Its television service starts for real in 1936, but is suspended when the war starts.

*1940s:* The BBC's television service returns in 1946. By 1950 it's broadcasting for 30 hours a week.

*1950s:* Thousands buy TVs to watch the Queen's coronation in 1953. In 1955, ITV goes on air, introducing TV ads and light entertainment shows.

## Audience share

For broadcasters, it's not always how many millions are watching a programme that matters – it's how big their share of the whole TV-watching audience is. In 2001, for the first time since commercial TV began in 1955, BBC1 got a bigger average audience share (26.8 per cent) than ITV (26.7 per cent).

*1960s:* BBC2 is launched in 1964. Colour TV is introduced.

*1970s:* TV takes over from the cinema as the nation's first choice in entertainment. By now, most families have one set and most programmes are in colour.

*1980s:* Channel 4 goes live in 1982, commissioning programmes instead of making them – independent producers go into business. Cable and satellite services make their debut. Breakfast TV starts. Video recorders go mainstream.

*1990s:* Low-cost, lightweight cameras take programme making out of the studios and onto the streets. Channel 5 (now called 'five') is launched in 1997. Digital television is born and interactive TV begins to take off.

## FEELING NOSTALGIC?

If you're yearning for the TV shows of yesteryear, check out your childhood favourites here:

- TV Cream (http://tv.cream.org) – golden oldie shows, theme tunes and ads
- Television Heaven (www.televisionheaven.co.uk) – hundreds of old programmes, indexed and reviewed.

## A NIGHT IN FRONT OF THE TELLY, 1950s STYLE

Ever complain that there is *nothing* on? Count yourself lucky you weren't watching in 1950. Here's what the schedule looked like on Sunday 9 July:

| | |
|---|---|
| 17:00 | For the Children – Annette in Fairyland |
| 17:30 | Children's Newsreel |
| 17:45 | Scenes from the Royal Military Tournament |
| 20:00 | Sunday Serenade with Vanessa Lee |
| 20:15 | The Orange Orchard – a 'Devonshire comedy' |
| 21:45 | The News (sound only) |

*(Source: BBC at www.bbc.co.uk)*

### Average audience share in 2001 (%)

| | | | |
|---|---|---|---|
| BBC1 | 26.8 | Channel 4 | 9.6 |
| BBC2 | 11.1 | Channel 5 | 5.7 |
| ITV | 26.7 | Cable and Satellite | 19.8 |

*(Source: ITC)*

# THE PRESENT

Britain is a TV nation. Watching TV is *the* most popular leisure activity. Ninety-nine per cent of us do it, for an average 25 hours a week. That's 3.6 hours a day. Almost all the UK's 24.6 million households have TVs.

Despite the take-off of multi-channel TV, the main terrestrial channels – BBC1, BBC2, ITV, Channel 4 and five – still get the biggest audiences. That may soon change, although many in the industry think digital TV was over-hyped in the late 1990s. Uptake has been slower than expected – 40 per cent of households have digital TV but, unless everyone has gone digital by 2010, the government will have to postpone plans to turn off the old analogue transmission signal.

A few years ago, media pundits were predicting the future of television would be in niche programmes on niche channels. With a choice of hundreds of channels, we would all be switching continually from one to the next. We would also be paying much more to watch TV. Some of that has happened, but the face of TV has not been totally transformed – yet. There are some successful niche channels, like Cbeebies, but many have failed to make an impact. We are channel hopping more than we used to, but we're also still quite lazy – if a channel can hook us in with one programme, it has a good chance of getting us to watch whatever's on next.

Although millions of us are paying for Sky subscriptions, and individually for football and films, there is still a definite reluctance to cough up for programmes we are used to getting for free.

## EVENT TV

Faced with the threat of multi-channel TV, and the prospect of twitchy-fingered viewers with a five-second attention span, the terrestrial channels have fought back, not just by launching their own digital channels, but also by locking viewers in with must-see, major event TV shows.

Think about *Who Wants to Be a Millionaire?*, *Big Brother* and *I'm a Celebrity...* – programmes that are shown every night for several weeks, and that have the 'water cooler' effect: everyone talks about them at work the next day. These big 'formats', that can be developed as 'brands' and sold abroad, are the holy grail in TV these days. There has also been a trend for 'theme' nights – '100 best' shows or issue specials – that lock viewers in for the evening.

## INTERACTIVE TV

Another weapon has been interactive TV, getting viewers involved by calling, texting, using their remote control or going online. 'The new interactivity is a powerful way to attract and keep an ever more promiscuous audience,' says Peter Bazalgette, chairman of Endemol UK.

He should know. Endemol makes *Big Brother* which, in its first two series, got 34 million phone votes. It was one of the first big shows to get viewers actively involved, texting their votes or navigating their own way round the house on E4.

## REALITY TV

*Big Brother* started another revolution too – reality TV. In the same way that docu-soaps dominated the schedules in the late 1990s, reality TV has been the defining genre of this decade, so far. Not only has it given viewers something different – real people – it has also been a boon for broadcasters because it is so cheap to make, costing £60,000–120,000 an hour, compared to £400,000 an hour for a comedy show and maybe half a million an hour for a high-quality TV drama.

## CHEAP TV

In general, programme-making has been getting cheaper. Ever since light-weight DV cameras gave programme-makers the freedom to go out and about filming, without the expense of a whole crew, costs have been dropping. Programmes can also be made in cheap virtual studios and have low cost special effects added. Producers making daytime TV shows now do it on amazingly small budgets.

## THE WORLD MARKET

British TV does well abroad. We are the second largest TV exporters in the world. It's not just shows made here that sell well, like *Mr Bean* or *Teletubbies*. Our ideas sell well too – formats like *Who Wants to Be a Millionaire?* and *The Weakest Link*. Even so, our TV exports market is *nothing* compared to America's. British shows make up 13 per cent of the TV exports market, but America's account for 68 per cent of it.

## THE RATINGS WAR

Whilst not all TV is entirely ratings-driven – for example with public service TV there are other factors at play – ratings are of great importance.

Nowadays, if a new programme doesn't attract a big audience fast, it often gets pulled or rescheduled. How are the ratings measured? By Barb, the Broadcasters Audience Research Board, which monitors the viewing habits of 10,000 people in 5,100 homes.

Everyone in the industry has their eyes on the 'overnights' – the ratings from the night before. To see them, check out TV Overnights in the Media section of Guardian Unlimited (www.mediaguardian.co.uk) – a great source of insider news and views.

It is not just technology that has been driving costs down. The recent, major downturn in advertising has meant all the broadcasters – except the licence fee funded DDC – have been feeling the pinch, cutting back on programme budgets and making redundancies.

Financial pressures also mean there is now more sponsorship of programmes and more merchandising. Broadcasters are continually looking for new ways to boost profits from programmes, whether it's through text voting or selling formats abroad. All in all, the TV business is getting more and more business-savvy.

> **❝**  THE STATE TV'S IN
>
> 'Don't lets be fooled by typical British self-deprecation. The reality is that British TV remains in rude creative health... [There are] great programmes we should all be proud of, from *Bloody Sunday* to *The Office*, *Pop Stars* to *Blue Planet*... British production ideas are transforming US entertainment, lifestyle and 'reality'. Britain has the lead in high quality international news, documentary and landmark series.'
>
> *Jana Bennett, Director of BBC Television*
>
> 'There's more choice than ever before and the sheer wealth of programming exposes the lazy canard of the detractors who claim there's "nothing to watch". Poppycock – the glass is half full and the Golden Age never was... TV is getting better.'
>
> *Peter Bazalgette, Chairman of Endemol UK, makers of* Big Brother, The Salon, Fame Academy *and more...*
>
> **❞**

## THE FUTURE

So what does the future hold? Since the dotcom bubble burst, there is a general reluctance in the media to make sweeping predictions for the future. But it doesn't take Mystic Meg to predict that TV is likely to get more business-focused, not less.

> The pressure to come up with big formats that can sell internationally is likely to increase, meaning there is a greater demand in the industry for people with sharp business skills.

Programme budgets will continue to fall, meaning producers will have to come up with ever-more inventive ways of making compelling TV on the cheap. Multi-channel TV will mean there is even more air time to fill, cheaply. Interactive TV will continue to grow, although who knows in what form. We'll all be watching digitally-transmitted TV although it doesn't look like we'll be saying goodbye to the 'big five' channels for quite a while.

As for the kind of programmes we'll be watching in future... Genres are always evolving (see Know Your Genres, from page 10). Reality TV, and all those lifestyle shows – DIY, property, cooking and gardening – will be replaced by whatever the 'next big thing' is. Broadcasters will probably continue chasing the treasured 16–34 audience, the people advertisers most want to reach – but it may also be that the UK's rapidly ageing population, including the rich baby boom generation, becomes the new target audience.

Who knows? If you do, you could have a very successful and lucrative career.

## What they think

'Is the future bright for a career in television? I think it's terrific.'

*Jon Snow*, Channel 4 News *presenter*

'We should all honour the past, steal what we can from it and then forget about it. Let's ignore those who claim that there's less talent than there used to be. Our industry is bursting with talent – if we've lost anything it's the ability to communicate with it and trust it, to give it a chance. TV used to give big creative opportunities regularly to people in their 20s and early 30s. We've got to start doing that again.'

*Mark Thompson, Chief Executive of Channel 4*

## ON THE WEB

The **IDEASFACTORY** website (www.channel4.com/ideasfactory; www.ideasfactory.com/film_tv/index.htm) offers an invaluable resource to young creatives looking to get in and get on in the UK's TV industry. It provides a wealth of inspiration and information in the form of fresh, challenging, original features, games, masterclasses, etc. Plus careers, business, training and funding information and services, as well as region-specific 'hubs'. **IDEAS**FACTORY users take advantage of the site's interactivity to create and post their personal profiles, small ads, and threads in the dynamic forums.

## KNOW YOUR GENRES

Programme genres used to be much more clearly defined. There were news programmes, arts programmes, light entertainment programmes and so on. Nowadays genres are more blurred – we have docu-soaps, entertainment news and reality dramas, a.k.a reality TV.

It is important to understand what the different genres in TV are, and how they are evolving. Thinking about them will help you get your head in gear to make the right career decisions and impress people at interviews – you will be expected to be clued up about what's what on TV.

## THE GENRES GAME

To give you an idea of how major broadcasters categorise different genres, check out the official BBC's 'genre structure'. At first it can seem a bit bewildering, but take a look at the different genres, and their sub-categories, and try and think of BBC programmes that belong in them. There are some examples below.

### The BBC's Genre Structure

1. **Drama.** Includes Drama Series (ongoing, indefinite series), Drama Serials (short series, that begin and end), Drama Singles (a one-off drama), Daytime Drama and Comedy Drama.
2. **Entertainment.** Sitcom, Comedy, Comedy Drama, Light Entertainment, Entertainment Formats (a specific format), Sport Entertainment, Daytime Entertainment, Factual Entertainment, Pop Music and Animation.
3. **Factual.** General Documentaries, Leisure, Factual Entertainment, Features, Daytime Factual, Events, Specialist Factual (Business, Natural History, Religion, History and Science), Arts and Culture, Current Affairs and Investigation.
4. **Programme acquisition.** Feature Films, Acquired Drama Series, Acquisitions.

### Examples

1. **Drama.** 'EastEnders' and 'Holby City' (Drama Series), 'Daniel Deronda' and 'State of Play' (Drama Serials), 'The Other Boleyn Girl' and the 'Lost Prince' (Drama Singles), 'Doctors' (Daytime Drama).
2. **Entertainment.** 'My Family' and 'Two Pints of Lager and a Packet of Crisps' (Sitcoms), 'Alistair McGowan's Big Impression' (Comedy), 'The Office' and 'The Royle Family' (Comedy Drama), 'Never Mind the Buzzcocks' and 'The Kumars at Number 42' (Entertainment Formats), 'They Think It's All Over' (Sport Entertainment), 'Top of the Pops' (Pop Music), 'Stressed Eric' (Animation).
3. **Factual.** 'Sahara with Michael Palin' (General Documentaries), 'Would Like to Meet' and 'What Not to Wear' (Leisure), 'Changing Rooms' and 'Ground Force' (Factual Entertainment), 'Bargain Hunt' and 'Kilroy'

(Daytime Factual), 'History of Britain', 'Walking with Cavemen' and 'Back to the Floor' (Specialist Factual), 'Walk on Dy' and 'Storyville' (Arts and Culture), 'Panorama' and 'Correspondent' (Current Affairs).
4. **Programme acquisition.** *Shrek*, *The Others* and *Amores Perros* (Feature Films), '24' (Acquired Drama Series).

KNOW YOUR SCHEDULES

If you watch TV carefully, you've probably already got a good idea of what kind of programmes get shown when. The art of scheduling – deciding which programmes to put in which spot – is crucial.

> The key to winning the ratings wars is not just coming up with winning programme ideas, but making sure those programmes are shown at the right time, in the right place.

Unless you become a scheduler, or one day a channel controller, you're unlikely to be making any decisions about what to show when. Even so, having a feel for the schedules and how they work is crucial for anyone who wants to work in TV. Employers will expect you to know the schedules and how they work.

Take a look at Channel 4's basic schedule. Jot down some Channel 4 programmes you know, and where you think they belong in the schedule. Also, have a good look through the TV listings to see what gets shown when, and if it fits with this basic schedule.

# C4 SCHEDULING

◼ LONG-RUNNING PROGRAMMES    ◼ COMMISSIONED PROGRAMMES

| | Monday | Tuesday | Wednesday | Thursday | Friday | Saturday | Sunday |
|---|---|---|---|---|---|---|---|
| 1800 | Friends rpt | Acquisitions | Acquired | Acquisitions | Acquisitions | | |
| 1830 | Hollyoaks | Hollyoaks | Drama | Hollyoaks | Hollyoaks | C4 News | |
| 1900 | C4 News | C4 News | C4 News | C4 News | C4 News | Repeat | |
| 1930 | | | | | Innovative | | C4 News |
| 2000 | Strong Narratives | Low Cost | Brookside | Brookside | Low Cost | Various | Landmark Series |
| 2030 | | | Ent Series | Lifestyle | Brookside | | |
| 2100 | Factual | Contemp. Factual | Docs | Contemp. Factual | Comedy Zone | Entertainment and Feature Films | Strong Narratives |
| 2130 | | | | | Comedy | | |
| 2200 | Films | Various | Docs | Various | US Acquired | | Films |
| 2230 | | | | | Comedy | | |
| 2300 | | Films | Entertainment | | | | |
| 2330 | | | | | | Films and Narrative Repeats | |
| 2400 | | | | | | | |

## ON THE WEB

Once you've got to grips with genres and schedules, have a laugh at them at TV Go Home, a cult comedy site, now also a book, that has a go at TV listings. See www.tvgohome.com. Another good, irreverent site for anyone who finds the Radio Times a bit traditional is Off The Telly, an online television fanzine. See www.offthetelly.co.uk.

## ON THE WEB

To find out more about TV genres, see the BBC's Commissioning website at www.bbc.co.uk/commissioning. It gives a great insight into how the BBC works, who's who, what it's looking for, what's hot and what's not, and how it goes about commissioning programmes.
Channel 4's site for producers, www.channel4.com/4producers is also very useful, with a bias towards the independent production sector.

'Be warned. There are not many cosy "jobs for life" in TV.'

# How many people are there and where do they work?

## HOW MANY PEOPLE WORK IN TV AND WHO DO THEY WORK FOR?

There are about 25,000 people working in broadcast television, according to the Skillset Employment Census 2002. That means 25,000 people working for one of the broadcasters – the companies that make and transmit TV. On top of that, there are about 13,500 people working in independent production – the 'indies' that make programmes for the broadcasters.

### THE BIG EMPLOYERS

- **BBC.** The BBC, the UK's biggest public service broadcaster, employs more than 26,000 people, but that includes a vast array of jobs, many of them unrelated to TV. About 14,500 work in programme-making.
- **ITV.** ITV is made up of 15 regional broadcasters and GMTV, the network breakfast show. Most of these regional companies are owned by two mega-broadcasters, Carlton and Granada. Granada has 5,000 employees. Carlton has 2,800.
- **Channel 4.** Channel 4 is a publicly owned publisher/broadcaster. It does not produce its own programmes but commissions them from more than 300 independent production companies across the UK. It does not,

therefore, employ programme makers. About 900 people work for it in other roles – schedulers, commissioners, engineers, press officers, etc.

- **Five.** Like Channel 4, five commissions its programmes from independents. About 250 people work for it.
- **BskyB.** BskyB is the king of the 'multichannel', digital sector. Its biggest service, Sky, has more than 6 million subscribers. About 9,000 people work for it, mainly in administrative, technical and engineering jobs.
- **ITN.** ITN does the news for ITV, Channel 4 and five, as well as running various other TV and radio news services. It employs about 750 people.
- **Independent production companies.** Around 25 per cent of BBC productions should come from independent companies. There are about 1,000 independent production companies. A handful are massive, employing hundreds of people. About 40 are what you'd call small to medium sized, employing 10 to 40 people. The vast majority are small, with less than ten staff, who hire freelances when they need them.

## THE BIG INDIES

These are the top ten independent production companies, as rated by their peers on the basis of profit, staff numbers and commissions.

1 Endemol UK
2 TWI
3 The TV Corporation
4 Talkback Thames
5 RDF
6 Lion
7 Tiger Aspect
8 Chrysalis
9 Wall to Wall
10 Princess.

(*Source:* Televisual *Magazine,*
September 2003)

## STAFFERS VS FREELANCES

Be warned. There are not many cosy 'jobs for life' in TV. It is estimated that about half the workforce are self-employed, a.k.a freelances. Lots of

# Freelancing – what it takes

To survive the freelance life, you'll need to be good at:

- selling yourself and your skills
- being flexible and working at short notice
- getting along with new people
- teaching yourself new skills, or blagging free training
- organising your life and your finances
- not taking it personally when you don't get work.

people are on short contracts, sometimes as short as a month. Two week contracts are not unheard of. That said, there is a lot more stability in TV than in other, similar sectors. In film, a staggering 98 per cent of people are freelance.

Permanent jobs tend to be with the broadcasters – just 19 per cent of people in broadcast TV are freelance, compared to 47 per cent in independent production, according to the Skillset Employment Census 2002. The people with the jobs are often in management and administration. Anyone working in broadcast engineering, studio operations and transmission is also likely to be in a job.

The big freelance areas are producing and production – about half the people working in these are freelance. There are then some specialist areas where most of the workforce is freelance; for example, 65 per cent of people working with cameras, 80 per cent of people working in costumes and 91 per cent of people in make-up and hairdressing are freelance.

All in all, there is no doubt that freelancing is a way of life for the TV industry.

## WHERE THE JOBS ARE

If you're looking for a profession that will let you live anywhere in the UK, this is not it. Forty-seven per cent of people working in the audio-visual industries are based in London, according to the Skillset Employment Census 2002. Another 15 per cent are near London, in the South East. The next biggest employment area is the North West, with 9 per cent of the workforce, followed by Scotland with 6 per cent and the West Midlands with 5 per cent.

The statistics may make it look more London-centric than it really is. There are many more opportunities for working outside London in TV than there are in other sectors of the audio-visual industries:

- The BBC has headquarters in Scotland (Glasgow), Wales (Cardiff) and Northern Ireland (Belfast), plus 11 other regional centres (Birmingham, Bristol, Leeds, London, Manchester, Newcastle, Norwich, Plymouth, Nottingham, Southampton and, for the South East, Tunbridge Wells).
- ITV has its regionally based operators – Anglia (Norwich), Border (Carlisle), Carlton Central (Birmingham), Carlton London, Carlton

Westcountry (Plymouth), Channel Television (Jersey), Grampian (Aberdeen), Granada (Manchester), HTV Wales (Cardiff), HTV West (Bristol), London Weekend Television, Meridian Broadcasting (Southhampton), Scottish Television (Glasgow), Tyne Tees (Newcastle upon Tyne), UTV (Belfast) and Yorkshire Television (Leeds).

- Channel 4 has a Nations and Regions department and an office in Glasgow. In Wales, 70 per cent of S4C's output is from Channel 4. Channel 4's licence requires it to commission 30 per cent of programmes outside London.

What you won't find so many of in the regions, are independent production companies. In fact, you won't find that many of them outside Soho – the epicentre of the TV and film worlds. That makes it hard for non-Londoners starting out in their careers. The capital is a tough city to turn up looking for work in, with sky-high rents and travel costs. And even when people have a few jobs under their belt, living the freelance life in London, with all its financial perils, goes on being hard.

'Even when people have a few jobs under their belt, living the freelance life in London, with all its financial perils, goes on being hard.'

## THE FACE OF THE WORKFORCE

So what does the TV workforce look like? At the top, it's still largely male, white and middle class. Or, if you want to be rude about it, 'male, pale and stale'.

But plenty of women are beginning to break through the glass ceiling – the BBC currently has women in its top jobs of Director of Television, Controller of BBC1 and Controller of BBC2. And further down the ladder, women are, in general, equally represented.

The same can't be said for ethnic minorities, although the outlook is a lot brighter in TV than other sectors of the audio-visual industries. Workers with disabilities are also seriously under-represented. Concerted efforts finally seem to be being made to improve their employment prospects.

As for the age of the workforce, well TV is a young(ish) man and woman's game.

### AGE

Walk into an independent production office and you won't see a lot of grey heads. The Skillset Freelance Survey 2000–1 found the average age of freelances was 39. However, the biggest age group in the workforce was 26–35.

There aren't statistics to show if it's the same story for people in jobs. The chances are that the average age of employees is higher, simply because there is more stability in staff jobs and people (in theory) can't get pushed out of them because of their age.

### WOMEN

Women rule in TV. Well, not quite, but they're on much more of an equal footing with men than in other professions, even within the audio-visual industries. Half the workforce in broadcast television is female, compared to a pathetic 15 per cent in film production, according to the Skillset Employment Census 2002.

When you look at different jobs, there are still some gender divides. Women are almost equally represented in producing, production and journalism, but very few work in technical jobs – in broadcast engineering, cameras, lighting and sound. In contrast, there are some jobs that are almost women-only – 93 per cent of people working in make-up and hair, and 82 per cent of people in wardrobe, are female.

Don't be put off if you're a female wanting to do one of the supposedly 'male' jobs, or if you're a man wanting to work in make-up or costume. You might have to put up with a few prejudices, but you may also be welcomed with open arms – trainers and recruiters are well aware of the gender imbalances in these roles.

In terms of life on the job, you're unlikely to come across extreme sexist behaviour in TV. It is, on the whole, a pretty civilised, female-friendly world, although there are always exceptions to the rule. You'll also find plenty of women who successfully manage to juggle their careers with having kids, although it's not always easy.

Freelances aren't entitled to maternity pay and it can be hard looking after kids during intensive periods of work, with long filming hours, often away from home, followed by lengthy sessions in the editing suite. If you work for a small production company, the working culture may be one where you're continually expected to give your all, family or no family.

## ETHNIC MINORITIES
Greg Dyke, the BBC's Director-General, once famously described the BBC as being 'hideously white', and BECTU, the biggest broadcasting union, recently branded television 'institutionally racist'.

There is no doubt that ethnic minorities are still under-represented in TV. Throughout the audio-visual industries, ethnic minorities make up 8.2 per cent of the workforce, according to the Skillset Employment Census 2002. This is a big improvement on previous years. However, 8.2 per cent doesn't look so good when you consider that 11 per cent of the UK's working population are ethnic minorities and that, in London, where so many TV jobs are, ethnic minorities make up 38 per cent of the general workforce, but only 12 per cent of the audio-visual industry workforce. Ethnic minorities do much better in public sector TV, where they make up 9.7 per cent of workers, compared to just 3.8 per cent in commercial TV.

An industry-wide initiative, called the Cultural Diversity Network, has been set up to increase the number of ethnic minorities working in broadcasting – see www.channel4.com/diversity. Its main backers are the BBC, ITV, Granada, Carlton, ITN, Channel 4, five and Sky.

## WORKERS WITH DISABILITIES
Very few people with disabilities work in the audio-visual industries – only 0.8 per cent of employees and 0.8 per cent of staff have disabilities, compared to 12 per cent of the general UK workforce. They are slightly better represented in TV and radio than they are in film.

All in all, the figures are depressingly low. However, training opportunities for people with disabilities are definitely improving and there is beginning to be greater awareness amongst employers about low employment levels.

This is largely thanks to the Broadcasting and Creative Industries Disability Network, founded in 1997, which encourages employers to employ more people with disabilities. It's backed by big employers, including the BBC, BskyB, Carlton, Five, Channel 4, Discovery Networks Europe, Granada and Turner Broadcasting System. You can read more about the BCIDN at www.efd.org.uk/www/guests/bdn.

## WHO'S IN DEMAND?

There is a serious shortage of people wanting to be TV presenters. No, not really... In the same way that everyone going into film wants to direct, a frightening number of people going into TV want to present. Ask a handful of hopefuls what it is they want to do in TV, and chances are most of them will probably say they want to be on screen, not behind it.

That doesn't mean there isn't a market for new presenters – everyone in TV is always on the look-out for fresh talent. Just because loads of people want to do it, it doesn't mean you can't, whatever this book or anyone says. If you think you've got what it takes, go for it.

But, bear in mind that there are hundreds of other jobs out there and that if you go for something less popular, your chances of getting in and getting ahead are much higher. While there are literally thousands of people wanting to be researchers, many having to work for slave wages to get their first break, there are great, not-badly paid jobs for people who want to be, for example, production accountants or transmission engineers.

These are some of the jobs identified by PACT, the Producers Alliance for Cinema and Television, and skillsformedia, the advice service run by Skillset, the Sector Skills Council for Broadcast, Film, Video and Interactive Media:

- Broadcast and transmission engineers
- Electrical engineers
- Production accountants
- Sales and marketing staff
- Carpenters
- Plasterers
- Scenic painters
- IT specialists

- Script editors
- Camera operators
- Production assistants.

## THE WORKPLACE

So what's it like from day to day? Ask any insider and the first thing they'll say is that it's not nearly as glamorous as you might think – don't confuse the glossy life you see on screen and in studios with working life behind the scenes.

And it's true. In general, most people working in TV are working in bog-standard, dreary looking open plan offices, not swanky advertising agency style ones with sofas, table football and beer on tap.

But... it's also true to say that a lot of people do have quite a nice time in TV. It's certainly not like being down a coal mine. Successful independent production companies can make a lot of money when they're producing big shows and some of them have office lives that reflect that – there are fresh flowers, champagne for staff birthdays and the occasional meal out in an expensive restaurant.

Often, though, the people getting those kinds of perks aren't getting the boring-but-important perks, like pensions, health insurance, and generous holiday and maternity rights. The flowers and champagne are a pay-off for working very long hours on short contracts.

Remember, too, that working environments vary dramatically, depending on what you do in TV. A camera assistant working on location for a drama

'Flowers and champagne are a pay-off for working very long hours on short contracts.'

series could end up working anywhere in the country, or world, doing 16-hour days, often in the freezing cold. A producer could spend months in the office, in pre-production, setting up a series, followed by months out and about with a crew filming it.

Others spend their entire working lives in the office. A video/film editor is pretty much guaranteed to spend his or her entire existence in a black, windowless editing suite.

**Jargon Buster:** *Production*

Programmes are made in three stages:

- *Pre-production* – the planning stage
- *Production* – the shooting stage
- *Post-production* – the editing stage.

### THE HOURS

There are lots of steady, 9–5-ish office jobs on the business side of TV, but if you work in programme making, life is unpredictable, to say the least. Although there can be quiet 'down' times in between projects, when the heat is on, it's stressful. Why? Because it takes a surprising amount of time to make a programme – what, for the viewer, is gone in a flash on screen, is usually the end product of hundreds of hours of research, planning, filming and editing.

It can be particularly stressful if you're working freelance or on short contracts at independent production companies that expect everyone to give 100 per cent all the time. The nature of the work can be stressful too. Researchers are often under intense pressure to find real people with extraordinary stories, willing to appear on TV at a moment's notice. Producers are often given tiny budgets to make a whole series for daytime TV, and only a few weeks to do it in.

So, for many people, TV is hard work. But the unpredictability of it is also what attracts a lot of people. If you think you can hack that kind of stress, and a life filled with extremely intense periods of work, it could be for you.

## THE MONEY

OK, cut to the chase – how much does everyone earn?

Working in TV won't make you a millionaire – unless you come up with a format idea like *Who Wants to be a Millionaire?* – but you won't be a pauper either. On the whole, people doing well in TV are handsomely paid. But there are also plenty of people starting out in TV who are very badly paid, usually by cheapskate production companies paying runners and researchers a pittance.

Piecing together a picture of average salaries is tricky. There can be big regional variations – people working in London tend to earn more, although that isn't always the case: one of the big advantages of working for the BBC is that its salary structure is nationwide, so a producer in London will earn the same as one in Glasgow. There can also be big variations from employer to employer.

**ON THE WEB**

To find out the minimum rates for freelancers, see the Freelance Production Agreement, agreed by BECTU, the Broadcasting, Entertainment, Cinematograph and Theatre Union, and PACT, the Producers Alliance for Cinema and Television, at www.bectu.org.uk/resources/agree.

> 'TV-types tend to be quite casual – few companies start before 10am and there are definitely no suits at work! But the freelance nature of the work can be unsettling emotionally and financially. TV is not something to get into for the money.'
>
> *Louise Doolan, Assistant Producer*

Another thing that makes it hard to get an accurate picture of pay is that in some jobs most people are freelance, so their annual earnings depend on how much work they get a year, whereas other jobs are staff, salaried jobs.

'The production environment can vary from being very relaxed and controlled to being as hectic as stock broking! One of the downsides is that it's freelance – once you're out of contract, you're out of money.'

*Michaela Furney, Researcher*

This should give you a rough idea of the kind of money we're talking:

*Programme Making and Journalism Jobs*

| | |
|---|---|
| Broadcast Assistant | £16,000–£25,000 |
| Production Assistant | £15,000–£30,000 |
| Researcher | £16,000–£35,000 |
| Script Editor | £20,000–£25,000 |
| Assistant Producer | £25,000–£35,000 |

# My experience – on the job

'I only work eight hours a day in my current job, which is really unusual for TV. I've worked for many different companies and normally the hours are long. I was sometimes working 22 hours a day.'

*Owen Matthews, Runner*

'I often work 19-hour days, sometimes without any break.'

*James MacDonald, Camera Operator*

'You negotiate your pay with every job. It's a bit like you're a car being hired. They call lots of people, ask what their rates are and then negotiate, depending on the number of days of work there are. The freelance rates may look like good pay, but multiplied by the number of days you work a year, it's not necessarily so good. Also, we're talking 12 hour days, so the hourly rate isn't actually that good.'

*Becky, Sound Recordist and Boom Operator*

| | |
|---|---|
| Floor Manager | £25,000–£40,000 |
| Reporter | £25,000–£45,000 |
| Producer | £30,000–£50,000 plus |
| Series Producer | £35,500–£55,000 plus |
| Director | £32,000–£50,000 plus |
| Presenter | £35,000–£100,000 plus |

*Technical and Artistic Jobs*

| | |
|---|---|
| Wardrobe Mistress/Master | £500–£1000 a week *(Freelance)* |
| Make-up Artist | £500–£1000 a week *(Freelance)* |
| Art Director | £600–£1,300 a week *(Freelance)* |
| Sound Technician | £15,000–£30,000 plus |
| Camera Operator | £500–£1,300 a week *(Freelance)* |
| Broadcast Engineer | £22,000–£50,000 plus |

*Business Jobs*

| | |
|---|---|
| Personal Assistant | £18,000–£25,000 |
| Schedules Assistant | £18,000–£25,000 |
| Advertising Sales Executive | £20,000–£30,000 |
| Press Officer | £25,000–£35,000 |
| Brand Manager | £25,000–£40,000 |
| Head of Acquisitions | £40,000–£60,000 |

There are opportunities for higher salaries in the industry. Mangement and Commissioning roles can often pay greater amounts.

## OTHER SECTORS

Loads of people working in TV work in other sectors too. They chop and change, maybe working on a TV show for a few weeks, a feature film for a couple of months, a commercial for a few days... It depends what you do, and whether you have a transferable skill. Here's a rough guide to those other sectors.

### FILM

Very few people work on feature films. The Skillset survey found just 1,500 people working on productions on its 2002 census day. Almost the whole workforce – 98 per cent – are freelance so many of them would not have been working on that day. All in all, Skillset reckons there are about 10,000 people who make a living in film and video production, working on and off throughout the year. There are no big employers, just hundreds of production companies, making feature films, short films and pop promos (music videos), who hire staff and crew in the run up to, and during, filming. The vast majority are based in London, although filming can take them anywhere.

## COMMERCIALS
There is a lot more money in making ads than there is in making films so, not surprisingly, there is a lot more work. Around 7,200 people work in commercials, according to the Skillset survey, and about 70 per cent of them are freelance. Because commercials are basically short films, lots of people who work on feature films make commercials when there is no film work about. Others do commercials full-time, preferring it because of the high budgets and often specialist techniques they get to use.

## ANIMATION
Just over 1,500 people work in Animation, according to the Skillset census. Many more are permanently employed than in film or commercials – just 40 per cent are freelance. They work on animation for everything from ads to feature films, children's TV to multimedia projects. Unlike film and commercials, there are jobs outside London: Bristol, Manchester, Cardiff, Edinburgh and Glasgow are all animation hot spots.

## SPECIAL AND VISUAL EFFECTS
Only about 300 people work in special, or visual, effects, according to the Skillset survey. In fact, there are probably about double that – people who work on effects, but are employed by broadcast, animation and post-production companies. Most work for small companies, in staff jobs.

## FACILITIES
The facilities sector employs about 8,000 people. Facilities companies, or houses, lease filming and editing equipment and post-production staff to everyone in the audio-visual industries, from big broadcasters to small indies. Of the 8,000 people in the sector, about 3,300 work in studio and equipment hire, mostly in kit maintenance and preparation, and about 4,600 work in post-production, editing programmes after filming has finished. Most companies are small, with fewer than 50 staff. About half the work is based in London.

## CORPORATE PRODUCTION

Corporate production is big business. There are around 2,500 corporate production companies in the UK with an annual turnover of about £2.7 billion. They work for companies, making corporate videos for training staff or sales purposes, staging live events, such as conferences or press days, and producing 'business TV'. About 3,200 people work in corporate production, mostly in staff jobs – only 16 per cent are freelance. Again, London is where much of the work is, although there are companies in Scotland, Wales and Northern Ireland too. Most employers are small to medium sized, with about 25 staff.

## INTERACTIVE TV

Of the 150,000 people working in the audio-visual industries on Skillset's 2002 census day, an amazing 46,000 were working in interactive media, many more than the 25,000 working in broadcast television. It's impossible to know how much of that work is TV-related, although interactive TV is certainly on the up. All in all, there are about 3,000 companies in the interactive media sector, usually small to medium sized, with less than 25 staff. Most people are in jobs, rather than freelance.

## A BEGINNER'S GUIDE TO INTERACTIVE TV

'Interactive Television' (iTV) gives viewers the ability to interact with their TVs. By using the remote control for their digital television, viewers can, for example:

- play along with a game show, e.g. *Who Wants to Be a Millionaire*?
- vote in a reality TV show such as *Big Brother*, or *Fame Academy*
- place a bet on a horse
- check cinema listings
- find out what the weather's like in Athens
- get the latest news
- test their knowledge with a quiz
- check their bank balance
- and much, much more...

Much of the content is provided by broadcasters. Broadcasters are excited about iTV as it gives them the opportunity to get closer to their audiences, providing additional information and entertainment.

Other content is provided by banks, shops, cinemas, newspapers and websites. For these content providers, Interactive TV is a great way to extend their brand and increase their customer base.

'There are too many wannabes who don't know what they wannabe. You need to have an idea, even a rough one, of the kind of direction you're heading in'

## 4. THE JOBS
# who does what

WHAT IT TAKES

Before we take a look at some of the jobs in the industry – at who does what – it's worth having a quick think about some of the personal qualities you will need to get ahead in TV.

Different jobs require different skills and personal attributes. What it takes to be a good camera operator is not necessarily what it takes to be a good producer. But there are some things all employers in TV are looking for, no matter what job you're applying for.

### WHAT TV BOSSES WANT
- *A passion for television* – for actively watching, not just for vegging-out viewing. For all kinds of programmes, in different genres.
- *A thirst for the industry* – an understanding of how the business works, what makes the schedules tick, what the current trends are, and what they're likely to be in future.
- *Opinions* – to have views about TV, about what does and doesn't work, what you do and don't like.

- *Imagination* – to be full of ideas and inspiration.
- *Life experience* – people who've been places and done things, people with interests and experiences that could help make good TV.
- *Reliability* – reliable, hard-working people who can be trusted to do what they're asked to, on time.
- *Initiative* – the confidence and drive to be able to get stuff done by yourself, without having your hand held too much.
- *Diplomacy* – team workers who can get along with other people on a project.

---

## Insider's tip

'The number one mistake people make is that they don't watch TV. It sounds ridiculous, but it's true. They think TV is something they want to work in, but they don't actually watch it, or not all aspects of it. They haven't looked through the schedules to see what the channels are all doing. They need to understand the way TV is broadcast, how the industry works. They also don't understand what the jobs they're going for actually involve. Worst of all, they turn up at production companies without having watched the company's content – and then they're surprised that we're offended.'

*Graham Stuart, Executive Producer of* V Graham Norton
*and Director of* So Television

---

If there is one thing guaranteed to make TV bosses groan, it's people who say they 'just want to work in television', without having any idea of what it is, exactly, they want to do. There are too many wannabes who don't know what they wannabe.

Of course, people do change direction during their career, but you need to have an idea, even a rough one, of the kind of direction you're heading in at first.

## MAIN ROLES

Let's look at some of the main roles – the usual suspects you'll find working in the TV industry:

- Producer
- Presenter
- Director
- Scriptwriter
- Broadcast Journalist
- Floor Manager and Assistant Director
- Camera Operator and Camera Supervisor
- Sound Technician
- Editor
- Broadcast Engineer
- Production Designer, Art Director and Set Designer.

### PRODUCER

It is the Producer who pulls a programme together, overseeing it from beginning to end. They come up with content ideas, plan and organise filming, hire and manage production staff and crew, write scripts and edit the finished tape (if it's not live). It's also up to them to make sure the programme stays within budget. In drama, they choose the writer and are then often intimately involved in developing and fine-tuning the script.

All in all, it's a big job, with a lot of responsibility and stress. It's also highly creative – the Producer is not just the decision maker, but the source of ideas for the programme. Most have some sort of specialism, in arts or science programmes. Others work across whole genres, in light entertainment or drama. News Producers are trained journalists who've opted to produce news programmes rather than be reporters. A Series Producer, or Editor, oversees several producers in the making of a whole series of programmes.

## ON THE WEB

Ever seen a TV show being recorded? If not, go – it's easy to get tickets to be in the audience. Check out Be On Screen (www.beonscreen.com) the BBC's What's On site (www.bbc.co.uk/whatson/tickets) and Channel 4's www.channel4.com/tickets.html, for information on what's being recorded when, and how to get free tickets.

> ❝ ON THE JOB – DRAMA PRODUCER
>
> 'I am creatively and financially responsible for the programmes I develop and make. My job is to see through programme ideas from initial concept, to script and commission and then employ a director and team to make the best drama possible within the budget. The great thing about my work is there is never an average day. If I am not in production I work from an office meeting writers, working on scripts, overseeing budgets and hopefully having brilliant ideas for the next programme. If I am shooting then my day starts about 7am when I come to the set for breakfast. The most important thing I do each morning is watch the rushes – everything we shot the day before. In the end everything depends on what we are getting on film.'
>
> *Francis Hopkinson, Drama Producer*, The Cops, A&E, The Jury ❞

## PRESENTER

We all know what Presenters do – they stand in front of the camera and present the programme. How much input they have varies. Some Presenters, on some programmes, have no involvement in planning the show but turn up on filming day, quickly scan the script and then, reading an autocue, speak the words to camera.

Some have a script but ad lib or interview guests as well. Others help research the programme, write the script and then deliver it. News Presenters, who are trained journalists, read a script but need to be clued up enough about the stories to go off-script and interview people and Correspondents live.

## DIRECTOR

Unless it's drama, it's Producers, not Directors, who are usually the big chiefs in TV. On some recorded shows, there may not be a Director at all, with the Producer overseeing the filming process. When there is one, it's

> **❝ MY EXPERIENCE**
>
> 'I broke into TV by accident. A friend from university was told to find an "unknown" so he panicked, like you would, and called up a whole load of friends. We had to traipse down to Canary Wharf to do 60-second audition tapes, pretending we all wanted to be Presenters. I got a call about eight months later from people working with Janet Street Porter and they asked if I could work on Live TV.'
>
> *Claudia Winkleman, Presenter,* Liquid News
>
> **❞**

their creative vision that shapes how the programme looks, what shots are used and so on. If it's a pre-recorded show, they direct the filming. If it's live, they stand in the studio 'gallery', with the Production Assistant and Vision Mixer, giving orders to everyone on the floor. They have overall responsibility for the set, how everything looks on camera and for managing the technical teams during filming or live broadcasting.

In drama, and a lot of comedy, the Director is much more the 'king' or 'queen' of the show. They can be involved in writing or editing the script and their direction of the actors and crew is crucial.

**SCRIPTWRITER**
The Scriptwriter, obviously, writes scripts. Sometimes that means working on already established series – soaps, ongoing drama series or sitcoms. Sometimes it means coming up with an original idea for a new series. That involves writing a proposal or a script, and then selling it to a Producer. Alternatively, they might get commissioned to write up someone else's idea, or to adapt a book or play for TV. Sometimes they get work rewriting other people's scripts. When filming begins, scriptwriters do not usually have much of a say in who gets cast and how their script is represented on screen, unless they have a close relationship with the Director and Producer.

" ON THE JOB – SCRIPTWRITER

'I love writing scripts. One of the things I love about it is that part of the art of it is trying to say a lot without using too many words. It's a very particular kind of writing, very sparse, more like writing poetry than prose. I also love the fact that the sky's the limit. You could write a script that makes you a multi-millionaire. Or, you could struggle all your working life to get anything made. It's definitely a job for dreamers. In terms of your day-to-day life, you're your own boss, you can work from home and organise your own time. If you're lucky, you do a mixture of working on your own and collaborating with others, going to meetings with Directors and Producers.'

*Amy Jenkins, Scriptwriter, including* This Life *and* Elephant Juice

"

" MY EXPERIENCE

'I started writing with a friend from university, Sam Bain, who's still my writing partner. We wrote an episode of an original sitcom idea. It took us ages – about 6 months – because we were both working in other jobs, just writing at the weekends. We sent it to about five agents, and one of them took us on.'

*Jesse Armstrong, Comedy Writer, including*
Peep Show *and* Smack the Pony

## ON THE WEB

The BBC gets sent 10,000 scripts a year, from established writers and newcomers. It promises to read at least the first ten pages of any script sent in. There is advice on what they are looking for, how long your script should be and where to send it on a BBC website called Writersroom – see www.bbc.co.uk/writersroom. Channel 4's producers site, www.channel4.com/4producers also contains information about how you can submit your ideas.

Some scriptwriters start out as script readers, reading scripts sent to production offices and writing up summary reports of them. They then move on to become Script Editors. This is also a route taken by Drama Producers. Others just start writing, or do a course – there are now some well-respected TV scriptwriting courses about. Established scriptwriters have agents who find them work and sell their ideas.

### BROADCAST JOURNALIST

Broadcast Journalist is a general title for journalists, including News Editors, Producers, Presenters, Correspondents (who cover a country or a particular area of the news, such as health or education) and general reporters. They all essentially do the same thing, whether it's in front of or behind the camera – they tell the news. That means researching, interviewing, writing, presenting, filming and editing. Some work on news bulletins, others work on news programmes. Most start out in radio and later move into TV.

> **"** MY EXPERIENCE – NEWS
>
> 'I've been able to travel the world seeing some dreadful situations, but I've also had the opportunity to meet some wonderful people, inspiring people and good people.'
>
> *Jonathan Dimbleby, News Presenter*
>
> 'This is my dream job, but initially it was very hard getting in. It's tough waiting one, two and then three years for the work you want and not having any money for ages.'
>
> *Asad Ahmad, Presenter and Reporter, BBC News* **"**

## FLOOR MANAGER AND ASSISTANT DIRECTOR

The Floor Manager manages the studio floor, listening to instructions from the Director in the gallery and relaying them to the crew, the Presenters, the guests and the audience. It is a bit like conducting an orchestra – they have to co-ordinate everything and everyone on the floor. Many come from a background in theatre stage management, others start out as Floor Assistants or Assistant Floor Managers.

If a programme is being made on location, the Floor Manager is known as Assistant Director (AD). On big productions, there will be a 1st, 2nd and 3rd AD. The AD, or 1st AD, is the Director's right-hand person, relaying instructions to the crew from the Director and keeping order on the set. They also write up the filming schedule and push everyone to keep to it – if things go wrong, or get delayed, it's their job to rethink the schedule.

## LIGHTING CAMERA OPERATOR AND CAMERA SUPERVISOR

The Lighting Camera Operator sets up and operates the camera. This can be in a studio, on location, or on the move. Camera equipment varies, depending on the job, and whether it's a single or multi-camera shoot. In studios, camera operators mostly use heavy electronic cameras. In news, lightweight cameras are the norm. While they may work alone in news and on other factual programmes, in studios there can be three to six cameras, mounted on pedestals and moved around during recording. They work using a camera script and listening to instructions in earpieces, given by the Director, via the Production Assistant.

When there is a team of lighting camera operators, they are co-ordinated by a Camera Supervisor. You'll also see credits for Steadicam Operators – these are Camera Operators using Steadicam, a piece of equipment, like a harness, that helps keep a hand-held camera steady, eliminating any unsteadiness in the operator's movement. Camera Operators tend to stick to working in either TV or film, rarely both.

# Beginner's guide to formats

TV is made using several different formats. Some big glossy dramas are still shot on film – the film stock used to make movies – but most TV shows are made using video or digital video. Digi-Beta (Digital Betacam) is the industry standard and the best quality, although DV (Digital Video) is increasingly popular, often with smaller crews on low budgets. For non-broadcast programmes, for training or corporate purposes, Hi8 and Mini DV can be used.

**❝** ON THE JOB – CAMERA OPERATOR

'Every job is different. One day I could be filming a pop star and the next I could be filming a woman taking her child into hospital. On average I'll get to the office for about 7 to check over all the equipment before heading off to a job. Once I'm there I'll have a chat with the production people to get an idea of what they want to do that day and then I'll just get on with it. If I'm interviewing talking heads, I'll scout around the location to find a nice place to film and set up, but it all really varies. I'm never office bound – I'm always out and about. I get to meet loads of interesting people that I would otherwise never have met, here and abroad. I end up in some very unusual situations.'

*James MacDonald, Camera Operator*

**❞**

## SOUND TECHNICIAN

Sound Technician is a catch-all title for various sound jobs. At the top of the tree are Sound Supervisors, with overall responsibility for the programme's sound. They oversee the setting up of microphones and then sit in the gallery with the Director and Vision Mixer, monitoring the sound coming from the mics and giving instructions to Boom Operators (the ones who holds mics in the air, on a big stick) in the studio. Then there are Sound Recordists, a.k.a Production Mixers, who record the sound during filming, either in the studio or on location, monitoring it through headphones and using a console to balance, boost and mix it.

## EDITOR

The Editor takes the recorded footage and cuts it together to make the finished programme, ready to be broadcast. So this is a post-production job. Together with the Director or Producer, the Editor decides which shots to use, in what order. When they're finished the sound gets tweaked, smoothed out and generally perfected by a Dubbing Mixer. In factual programming, Editors usually work unassisted. In drama, they have an

Assistant Editor to digitise the tapes (load them into the computer) and keep records.

---

# Jargon Buster: *Offline and Online*

'Offline Editing' refers to the quality of the pictures during this particular stage of the editing process, which are lower than during 'Online Editing'. Avid is the trade name of one of the most commonly used editing systems (both Offline and Online – often the same kit but with a 'dongle', a plug-in hardware component, to unlock certain features).

During the offline edit an EDL (Edit Decision List) is produced which lists all the shots that are going to be used (by their In and Out timecodes i.e. where the shot begins and ends, accurate to frame). This list is then used to quickly edit on a more expensive, broadcast-quality online system.

Editing software is getting more sophisticated and has become less hardware dependent (although you can never have too much processing power, speed and memory). A few years ago Apple released FinalCutPro to wide acclaim and that offers real-time, online quality without any extra hardware.

There are dozens of video editing packages out there.
Some (e.g. Media100 and Lightworks) are hardware dependent – they have their own video capture cards, which determine the amount of compression. Others (e.g. FinalCut and Premiere) can work on any computer with an adequate spec.

---

## BROADCAST ENGINEER

The Broadcast Engineer, broadly speaking, designs, installs, maintains and operates broadcast equipment, in studios, on locations, working on recording, editing and transmission equipment. There are lots of specialist roles under the broadcasting engineering umbrella – Studio Engineers, who look after all the studio equipment; Transmission Engineers, who test and maintain transmission equipment; and Post-production Engineers, who oversee all the equipment used during post-production. There are others too – and they're all in high demand.

> **ON THE JOB – BROADCAST ENGINEER**
>
> 'Primarily my job is to keep the digital platform on air at all times. Secondary is to maintain all the broadcast equipment. My day at work normally starts at 7 a.m. and finishes at 7 p.m., and I work seven days out of every fortnight. We man the platform 24hrs a day so there are also some night shifts. All channels are monitored in our Network control area and as soon as an "off air alarm" sounds, reporting that a channel is off air, either by losing audio or video or both, we rush to our master matrix and try to find the faulty piece of equipment. When we are not dealing with off-air problems we are fixing the broken equipment on the repair bench.'
>
> *Terry Sargeant, Broadcast Engineer, Sky*

## PRODUCTION DESIGNER, ART DIRECTOR AND SET DESIGNER

The Set Designer, Production Designer and Art Director design and oversee the creation of the set. On some productions there is a Production Designer and Art Director, the second-in-command. On others, the job title is Set Designer.

> **MY EXPERIENCE – PRODUCTION DESIGNER**
>
> 'I love the diversity, the speed at which you have to come up with ideas and turn them around, the studio building and the final result.'
>
> *Alison Dominitz, Production Designer*

They work closely with the Producer and Director in deciding how the programme should look and then instruct everyone in the art department (Prop Makers, Set Dressers and Graphic Artists) on what they want done. They also work closely with the crafts team (the Construction Manager, Decorators, Carpenters and Plasterers), giving them directions on how they want the set built.

## ENTRY-LEVEL JOBS

What about the jobs at the other, bottom end of the ladder? Where did the people in the big jobs start out? Here are some of the entry-level jobs in TV:

- Runner
- Production Secretary
- Production Assistant
- Researcher
- Camera Assistant
- Sound Assistant
- Lighting Assistant
- Assistant Editor
- Wardrobe Assistant
- Make-up and Hair Assistant
- Art Assistant
- Script Reader
- Assistant Location Manager
- Casting Assistant.

### RUNNER
A Runner… runs around, fetching and carrying whatever they are asked to fetch and carry. They work in the production office, running errands,

> 'A Runner… runs around, fetching and carrying whatever they are asked to fetch and carry.'

delivering tapes and messages, making tea, but also doing organisational and research work. They can also spend time on set or location – fetching props, assisting the Camera Operator, Sound Recordist or anyone else who needs something done.

## Getting in

You don't need qualifications or training to become a Runner, but you might need them if you want to progress into other jobs. Lots of Runners have degrees, and postgraduate training too. Loads of people start out as Runners, before going on to specialise – in a production or technical role, say. Many get their first break by writing to production companies. It helps if you've done work experience or worked on student productions.

"

ON THE JOB – RUNNER

'I do a bit of everything. I run the dispatch, I work in the post room, I dub video cassettes, I sometimes drive the CEO's car. I also do a bit of camera work – recording auditions and I set up technical equipment. I'll even fix the plumbing in the offices if that goes wrong. I work in a lovely office, with about 200 people. I'm one of the six office Runners. I don't go out of the office that much.'

*Owen Matthews, Runner*

'I got my first job as a fill-in – one day covering for another Runner at a pop promos company. I then became their regular fill-in, running around Soho doing errands, and then – eventually – their Runner. It can all seem like a long haul, but there are people who start running in their early 20s and are Producers by their late 20s.'

*Karen, Runner*

Working as a Runner gives you a valuable insight into the industry and an overview of all aspects of working in it. Most people start off as a Runner, it provides a foot in the door and the amount of experience gained is incomparable. It is one of the best springboards to all other jobs in the industry.

## PRODUCTION SECRETARY

Production Secretaries work in production offices, giving secretarial and administrative support to the production team – the producer, Production Manager, Production Co-ordinator, Production Assistant, Researcher, etc. They run the office, type scripts, make travel arrangements and do any other admin or organisational work necessary.

### Getting in

You'll need basic computer and typing skills. Jobs do get advertised, although some people get jobs through agencies and by temping. Traditionally, especially in the BBC, secretaries have been able to move up to being Production Assistants, Researchers or other roles, but people can also get stuck and pigeon-holed as admin, rather than creative, types.

## PRODUCTION ASSISTANT

Being a Production Assistant isn't necessarily always an entry-level job. Lots of people are Production Secretaries first. But in small companies, or on small productions, you might get lucky.

Production Assistants assist the Producer and Director, in the production office and in the studio or on location. In the office, they do organisational and administrative work – booking studios and equipment crews, writing up and distributing scripts and schedules. During filming, they sit with the Director in the gallery, cueing in Camera Operators and Technicians on the 'talk-back' system and timing the programme to make sure it comes out at the correct length. If working on location, they make sure the script is kept to and time shots. They are sometimes called Script Supervisors.

## Vital stat

There are 1,300 Runners working in film and TV in the UK, according to Skillset. Forty-six per cent of them are freelance, 52 per cent are women and 5.8 per cent are ethnic minorities.

*(Source: Skillset Employment Census 2002)*

## Getting in

You don't need a degree, but many have them and entry is competitive. IT and secretarial skills will help. There are a few training opportunities – see the FT2 New Entrant scheme in Chapter 6, page 75. Some start out as Production Secretaries and then become Production Assistants. Some go on to other studio jobs, as Floor Managers or Vision Mixers, while others progress in the production office, becoming Production Co-ordinators and Managers.

## RESEARCHER

Again, this isn't necessarily a first job, although it can be. Many Researchers stay in the role for years, becoming well-paid, highly experienced Senior Researchers. What do they do? Help brainstorm ideas for programmes; chase up story ideas; do background research, in libraries, archives and online; find possible guests and interviewees for programmes; go on 'recces' to check out interviewees and locations; write

---

**❝**     ON THE JOB – RESEARCHER

'A large part of the job involves finding suitable contributors to take part in the programmes that I am working on. This can mean anything from cold calling companies and organisations, to research on the internet. Once suitable people have been found, meetings are usually set up to visit the individuals in person and get them on tape, so that the APs and Producers can view them back in the office. Character profiles are usually written on each individual so that the Producers can get a feel for the kind of people they are and their suitability for the programme. The rest of the job usually involves helping to set up shoot days and acting as a liaison between the contributors and the Producers. Assisting on shoot days normally involves looking after contributors, taking continuity notes and logging shots. Sometimes DV filming is required.'

*Aislinn McIvor, Researcher*

briefing notes and questions for presenters. During recording, they may also be there to look after guests and make sure the presenter has everything he or she needs.

In news, Researchers are often called Broadcast Assistants and are junior journalists. They work in the newsroom, doing research, checking facts, making calls, setting up interviews, doing interviews and writing briefing notes for Presenters.

## Getting in
Most Researchers have degrees. Those working in news may also be trained journalists. There aren't many training opportunities – see the FT2 Researcher Training Scheme in Chapter 6, page 76 – although some companies take on trainee researchers. Some get their first break as Production Secretaries or Runners, and then move up to Researcher. Many move on to become Assistant Producers and then Producers.

## CAMERA ASSISTANT
A Camera Assistant assists the Lighting Camera Operator, looking after the tapes, making sure there are enough of them, labelling them and keeping records of what has been recorded on each tape. They also help move camera equipment, rig monitors and move cables out of the way as cameras move around the studio. Camera Assistants are also known as Clapper Loaders, but that's usually when working in film.

## Getting in
There are some training schemes for new entrants to camerawork provided by FT2 and the BBC, for example (see Chapter 6, from page 73). Some people manage, through letter writing, work experience or contacts to get freelance work as a Camera Assistant and then learn on the job. It will help if you've done some filming yourself, on your own or student projects. If you want to progress from Camera Assistant to Lighting Camera Operator, you will definitely need training.

## SOUND ASSISTANT
On big productions, the Sound Assistant helps the Sound Recordist and Boom Operator record the sound – helping move and set up equipment, positioning microphones, checking leads and cables. On smaller scale programmes, they may operate the boom.

## Getting in

There are no qualifications you *have* to have to be a Sound Assistant, but it helps if you've done some physics and maths. For one of the training schemes available – FT2 – practical experience in wiring and soldering is an advantage. The BBC also offers training (see Chapter 6, page 73). Some people get experience working with sound in hospital or local radio, as well as volunteering to do the sound on student films. The basic career route is to become a Sound Assistant, then a Boom Operator and then a Sound Recordist.

## LIGHTING ASSISTANT

The Lighting Assistant works with the lighting team, helping transport and lift equipment, set it up and maintain it during recording. They are generally on hand to help out the lighting team which, depending on the size of the production, can include a Lighting Director, Gaffer, Lighting Camera Operator and Best Boy.

## Getting in

Most lighting technicians learn and train on the job. However, almost everyone coming into the profession is already a qualified/apprenticed electrician. The BBC has some training opportunities.

## ASSISTANT EDITOR

The Assistant Editor helps the Editor in the cutting room. They organise the material the Editor needs, loading tapes into the machines, digitising it so that it can be edited on computer systems and keeping records of where everything is. They may also do some assembly editing on behalf of the Editor. On big drama productions, there can be more than one assistant. But often, on factual programmes, there is no assistant – the Editor works alone.

## Getting in

Lots of Editors, and Assistant Editors, start out as Runners on post-production, or facilities houses. They then graduate to become Assistant Editors. There are some training opportunities for new entrants – see FT2 in Chapter 6, page 76.

## WARDROBE ASSISTANT

Wardrobe Assistants are junior members of the Costume Department, a.k.a Wardrobe. They help alter and look after costumes, dress performers and generally run errands. On some TV programmes, the role of wardrobe is much more basic – to look after the Presenter's clothes, making sure they're ready, clean and ironed, for the show.

## Getting in
You don't need an art and design or fashion degree to work in wardrobe, unless you want to become a Costume Designer. If you don't have one, you will need qualifications or experience in sewing, altering and maintaining clothes. There are some training schemes, like FT2 and the BBC's Design Vision scheme (see Chapter 6, page 74).

## MAKE-UP AND HAIR ASSISTANT
Make-up and Hair Assistants work in the make-up room, assisting the Make-up Artist/s. They can also work on set. Jobs vary enormously. On some studio programmes, doing the make-up may just be a question of applying make-up to Presenters and performers to make them look 'normal' on screen. For dramas, the make-up can be complex, with prosthetics and specialist wigs. Assistants may help research make-up design, apply make-up and cut hair.

## Getting in
You will need to have done a course in beauty therapy and make-up. Some lead to the Hairdressing and Beauty Industry Authority NVQ/SVQ in Beauty Therapy Levels 2, 3 and 4. There are some training opportunities, for example with FT2 and the BBC's Vision design scheme – see Chapter 6, page 74. Senior, specialist Make-up Artists often have a degree in art and design.

## ART ASSISTANT
On big productions, mostly dramas, there may be an Art Department Assistant, or Runner, who helps the Production Designer and Art Director. Sometimes their duties are basic – making tea, running errands, ordering supplies. They can also be asked to do set drawings, copy and distribute plans, do period research, help find or make props and graphics and help dress the set.

## Getting in
Almost everyone has an art and design degree, maybe specifically in theatre design, interior design or architecture. There are some training opportunities, with FT2 and the BBC's Vision design scheme – see Chapter

6, page 74. Others get their first break freelancing as Assistants and then progress to become Assistant Art Directors or Design Assistants, Art Directors and maybe Production Designers.

> **ON THE JOB – ART DEPARTMENT ASSISTANT**
>
> 'I buy tea bags, milk, sugar – all that nonsense. But I also get to measure up and draw into plan form all the locations, make models, design graphics (logos, letters, signs for fictional establishments, etc). I also draw up details of sets for the construction team to build.'
>
> *Phil, Art Department Assistant*

## SCRIPT READER

Script Readers plough through the hundreds of scripts that get sent to production offices and assess whether or not they are worth being considered by the producers. They read the script, and then write a report on it. Many do this freelance, getting paid per script.

### Getting in

A lucky few might get jobs doing it, but you're more likely to get experience as a freelance Script Reader and then, later, a job as a Script Editor. Finding script reading work is a matter of luck – people use contacts or write to production offices to ask if there is any work going.

## ASSISTANT LOCATION MANAGER

The Assistant Location Manager, a.k.a the Location Assistant, does some of the leg work for the Location Manager – the person responsible for finding and organising locations (inside and outside) to film on. They go out and about looking for suitable locations and then help organise permission and fees for filming there.

## Getting in

There are no relevant formal qualifications or training. FT2 has a trainee scheme for Assistant Location Managers – see Chapter 6, page 76. Also, check out the Guild of Location Managers – see Chapter 8, page 100. Most start as Runners, become Scouts, Assistant Location Managers and then Location Managers.

## CASTING ASSISTANT

Casting Assistants work for Casting Directors, doing administrative and organisational work – making calls, sending faxes, opening letters, speaking to actors' agents and organising times for them to come in to do readings. They also do 'talent scouting' work for the Casting Director, going to see new actors in drama school or theatre productions, and reporting back on who might be worth auditioning.

## Getting in

There are no set routes in, or specific training. More important is a good knowledge of actors, a memory for faces and a passion for drama. Some start as runners on drama productions, then move into casting. Casting Assistants can also double up as PAs, doing admin or reception work, so getting a job as a PA for a Casting Director can be a way in.

## ENTRY-LEVEL BUSINESS ROLES

There are dozens of business-related roles in TV, and many of them offer entry-level opportunities as, for example, a Planning Assistant in Scheduling (someone who helps plan the schedule), an Acquisitions Assistant (working in Programme Distribution), a Press Assistant (working in Public Relations), a Legal or Rights Assistant (working on legal and rights issues), or a Marketing and Sales assistant. Getting in is, essentially, down to looking out for job ads that specify entry-level positions, or taking the initiative to write to different departments in different organisations to ask if there are opportunities.

---

**ON THE WEB**

If you want to find out more about where programmes and films get shot around the UK, have a look at the Movie Map, a guide to some of the most popular film locations in the country. See Movie Map at www.visitbritain.com/moviemap.

# OTHER JOBS... IN A SENTENCE

*Channel Controller* – chief of a channel, the person who ultimately shapes the content and schedule of a channel, deciding what gets shown when.

*Commissioning Editor* – commissions programme-makers and independent producers to make programmes, usually within a particular genre – drama, comedy, factual entertainment, etc.

*Scheduler* – works with Channel Controllers and others in shaping and planning the schedule, in the short and long term, liaising with Commissioning Editors and programme acquisition staff.

*Assistant Producer (AP)* – the Producer's right-hand man or woman, assisting in researching and developing the programme and communicating the Producer's instructions to the production crew.

*Development Producer* – develops new ideas for programmes, thinking them up, researching them and writing them up as proposals to pitch to Commissioning Editors.

*Production Manager* – oversees the budget and the administrative side of production, booking the cast, crew and equipment and organising filming and editing schedules.

*Production Co-ordinator* – co-ordinates the production office, overseeing paperwork (progress reports for the Ddirector, call sheets for crew and performers, script revisions); also books equipment, transport, etc.

*Production Accountant* – manages a programme's finances, works out budgets, keeps the books, oversees the cash flow and does any other financial work required.

*Location Manager* – reads the script and finds suitable locations, getting clearance from owners, councils, police, etc to film there, negotiating fees and arranging facilities for the crew.

*Casting Director* – organises the casting process for dramas, working closely with the Director and Producer to secure the right actors for the right fee.

*Script Editor* – develops and edits scripts with writers, but also looks for new script-writing talent and new ideas for projects; a junior Drama Producer.

*Vision Mixer* – works in the gallery, carrying out the director's instructions, cutting and mixing between different cameras or video sources, adding effects, captions and graphics as the programme is recorded or transmitted.

*Dubbing Mixer* – mixes, smoothes out and perfects the sound on recorded programmes.

*Focus Puller* – assists the Camera Operator, making sure the shot stays in focus, changing lenses, adjusting the focus during the shot; is also responsible for the maintenance of the cameras.

*Rigger* – a member of the camera team, laying out cables and the equipment that moves and lifts cameras during filming (tracks, cranes and dollies).

*Grip* – another member of the camera team, operating the equipment that moves and lifts cameras during filming (tracks, cranes and dollies); if a Key Grip, manages the other Grips.

*Lighting Director* – designs and oversees the lighting, both in studios and on locations.

*Gaffer* – the chief electrician on the set, responsible for setting up lighting and managing other electricians on the set.

*Best Boy or Girl* – an electrician and assistant to the Gaffer, helping set up and maintain lighting equipment.

*Construction Manager* – heads the set construction team (Carpenters, Painters and Decorators), working closely with the Production or Set Designer to make sure everything is built how it was designed.

*Carpenters, Painters and Plasterers* – do any carpentry, decorating and plastering work required on the set, working under the Construction Manager.

*Property Master or Mistress* – in charge of researching, finding and managing all the props used by the performers.

*Graphic Designer* – designs opening titles and credits, but also graphics used during programmes, and graphics for props, signs and anything else that appears on camera, including some visual effects.

*Costume Designer* – works on big drama productions, researching, designing, finding and making costumes.

*Make-up and Hair Artist or Designer* – applies make-up and arranges hair for Presenters, performers and members of the audience; for drama, they research and design make-up and hair, and are sometimes experts in specialist wigs and prosthetic special effects.

*Press Officer* – works for a channel or an agency, running press campaign to raise the profile of programmes, and answering queries from journalists.

Have a look at the Skillsformedia website, www.skillsformedia.com/getting_in/resources/the_business/gettingin_busi ness_jobs_2.htm, for in depth information about the jobs in the industry.

## THE RIGHT PERSON FOR THE JOB

What are you like? What job are you cut out for? Before you get carried away with daydreams about directing, fantasies about floor managing or visions of yourself as a Vision Mixer, see if any of these sound a bit like you.

1. **You have loads of ideas**. You're full of them. Everywhere you go, everything you see, you pick up ideas. You're always on the look-out for the next big thing. You love talking to people, asking them questions, finding out what's going on. You're also a doer – you're really organised and you love working on projects with other people. Even when you're very busy, you're calm.

2. **You love TV**. You watch all the soaps, all the 'Pop Idol' spin-offs and anything else that's on the box in prime time. You'd do anything to work in such a glamorous industry. You always read *Heat*. You know everything about TV celebs, what they're wearing and who they're going out with. You're also very sociable. You love meeting people and going to parties. You also love travelling. You want to see the world.

3. **You're a born problem-solver**. You're great at creating calm order out of a chaotic mess. You like fiddling around with things for hours on end, often on your own. You are methodical, but you're also creative – you

have a natural sense of pace, of rhythm, of how things should flow. You're also very diplomatic – you're good at gently suggesting solutions, but don't mind if your idea gets vetoed.

4. **You are an oasis of calm**. You have great co-ordination and can do ten things at the same time, concentrating for long periods. You're a perfectionist – highly accurate – and love anything technical. But you're also a creative soul. You think about how things look and have a sharp eye for detail. You don't have a problem taking orders, but can also give them yourself. You thrive on team work.

5. **You're nosy**. You like finding things out, asking people questions. You are confident and have no qualms about calling complete strangers and asking them to tell you everything they know about something you know nothing about. You're a fast learner, and can absorb lots of facts and information. You are also charming enough to be able to persuade people to do things they really don't want to do. You have common sense, good judgement and you're very organised.

6. **You're a rock**. You are the most reliable person you know. Everyone trusts you. You can perform under pressure, receiving instructions and giving them at the same time. You're well co-ordinated and cool as a cucumber in a crisis. You have the authority (and a loud enough voice) to get people to obey you, but you're not a dictator – you can calmly persuade people to do things, now!

## IF THAT'S YOU...

1. How about being a **Producer**? They are hard-working and multi-skilled – they can think big ideas while keeping their eye on small details (like budgets). They're full of ideas for new programmes, and new ways of doing things, and good at inspiring other people, getting them enthused about a project. They are also very practical and good at solving problems. When a guest doesn't turn up for a live show, or they realise they haven't got a crucial shot for a programme they don't panic – they come up with a solution.
2. Sounds like you'd like to be the next Cat Deeley or Vernon Kay. And you could be, but... loving TV and the whole celeb thing is not enough. Only a few people have what it takes to be a **Presenter**. You need looks that work on camera; the nerve to do what can be a terrifying job, especially live; the ability to read from an autocue; and the confidence to ad lib in between.

'You may get to travel, but you won't see
much of anything – you'll be too busy working.'

If you can do all that, it's a great job, but not nearly as glamorous as it looks on screen. You may get to travel, but you won't see much of anything – you'll be too busy working. Presenters also have a short shelf life, so you could be unemployed at 30.

3. Maybe you should think about being an **Editor**. Editors need to love messing around with new equipment and software. But they also need a strong creative streak. They're not just technicians – they play a crucial role in shaping a programme and its feel. They need to be able to help the Producer or Director achieve the programme they dreamt of from the rushes they ended up with, fixing thousands of problems along the way, sometimes holding their hand while they do it.

4. Looks like you might make a good **Lighting Camera Operator** or, at first, a Camera Assistant. Lighting Camera Operators are technically proficient, but also need to be creative – on some jobs they just point the camera straight at the presenter and stay there, but on others they need to create a look, using style and imagination. Lighting Camera Operators do have status, but don't always get the praise they're due – the Director takes that.

5. Well, you've got what it takes to be a **Researcher**. A good Researcher can be given the vaguest of briefs – for, say, a programme about single people – but can then go away and find out everything there is to know about being single, exploring every possible angle, finding all the facts and figures, as well as 20 people, all good speakers, to go on air and talk about it, even though they don't want to.
     They are great communicators – after doing their research, they can convey the important bits back to the Producer and Presenter. And on top of all that, they are highly organised, making notes, arranging interviews, crews and anything else that needs doing.

6. You'd be an ideal **Floor Manager** or **Assistant Director**. As a Floor Manager, you are the Director's voice in the studio, carrying out his or her instructions, getting everyone on the floor – from Camera Operators to Presenters – to do what they are meant to do, at the right time. As an

Assistant Director, you're the Director's voice on set, on location. You're responsible for the smooth running of the whole operation, for keeping discipline but also being tactful enough to know when people working for you are fed up, and why.

## HOW THEY GOT THERE

There are no set career paths in TV – different people take different routes, in different jobs. But here are some of the ways people get to the top of the tree:

*Lighting Camera Operator* – Runner → Camera Assistant, Focus Puller → Camera Operator → Camera Supervisor → Lighting Camera Operator

*Floor Manager* – Theatre Stage Manager → Floor Assistant → Assistant Floor Manager → Floor Manager

*Editor* – Post-production Runner → Assistant Editor → Editor

*Sound Supervisor* – Runner → Sound Assistant → Boom Operator → Sound Recordist, Mixer → Sound Supervisor

*Set Designer* – Art and Design graduate → Art Department Runner, Assistant→ Assistant Art Director → Set Designer, Art Director, Production Designer

*Producer – Factual and Entertainment* – Runner, Journalism Trainee → Production Assistant, Researcher → Assistant Producer → Producer

*Producer – Drama* – Script Reader → Script Editor → Producer

*Director – Factual and Entertainment* – Runner → Researcher → Assistant Producer, Producer → Director
or
Runner → Camera Assistant → Lighting Camera Operator → Director

*Director – Drama* – Scriptwriter, Producer, Lighting Director, Assistant Director, Actor → Director

*Presenter* – Runner, Researcher, Reporter, Actor or Model → Presenter

*Correspondent* – Trained journalist → Radio Reporter → TV Reporter → Correspondent

'Think about what kind of working environment you want to work in and what kind of work would best suit your talents.'

# different types of job for different types of people

If you know you want to work in TV, but still don't know which job is for you, maybe you need to think more generally about which area of TV you would like to be in – what kind of working environment you want to work in and what kind of work would best suit your talents.

Do you love doing anything arty? Or have you got a good head for business? Do you have strong techie tendencies? Or are you a news junkie? Maybe you're a natural born doer, or a compulsively creative writer...

If you're still not sure, take a look at the different types of jobs for different types of people:

- Doers and makers
- Creators
- Artists
- Techies
- Business heads
- News hounds.

# DOERS AND MAKERS

If you like working on projects, seeing an idea through from beginning to end, making and creating something out of nothing, being both creative and sometimes technical, but always ultra-organised, then maybe a job in producing or production is for you.

## THE JOBS
Producer, Director, Assistant Producer, Production Manager, Location Manager, Assistant Location Manager, Casting Director, Casting Assistant, Production Co-ordinator, Production Assistant/Script Supervisor, Floor Manager, Assistant Director, Researcher, Production Secretary, Runner.

## THE LOW-DOWN
When people in the industry talk about producing, they're talking about Producers. When they talk about production, they are talking about all the production jobs under the Producer – from Director to Location Manager to Runner. All in all, there are about 30,600 people working in producing and production roles in the audio-visual industries, the majority of them in TV.

All of these jobs are, some of the time, office based, some more so than others. But nearly all of them also require time out of the office – scouting around, doing interviews, being on location or in the studio.

Almost everyone starts out at the bottom – as a Runner, Production Secretary, Production Assistant or Researcher – and works their way up. Some people have media-related NVQs, degrees and postgraduate qualifications, but they are not a must. There are short courses in Production and Skillset NVQs that can be done once you're working.

> 'Almost everyone starts out at the bottom – as a Runner, Production Secretary, Production Assistant or Researcher – and works their way up.'

**WHAT YOU NEED TO BE**
Full of ideas. Organised. A good talker and listener, with strong interpersonal skills. Tactful and able to work under pressure. For more senior roles, a born leader.

**WHAT YOU NEED TO HAVE**
Research, organisational and administrative skills.

CREATORS

If you've got a great imagination and you love dreaming up stories, thinking up characters and observing how people talk, maybe you should write.

**THE JOBS**
Scriptwriter, Script Editor, Script Reader.

**THE LOW-DOWN**
It is hard to know how many Scriptwriters there are. The Writers' Guild of Great Britain, an organisation representing professional writers, says two-thirds of its 2,000 members are TV and film Scriptwriters, but there must be thousands more out there, already working in the industry or writing away at home, hoping to get in. Some Scriptwriters only write for TV, others write for TV and film. Some write novels and work as journalists too.

Unless you have a writing partner, which comedy writers often do, writing is a solitary business – you're at home, alone, and it's up to you to be motivated enough to get up every morning and start working. If you're lucky, and you get your scripts considered and even made, life gets a bit less lonely and you start going to meetings with Producers. If you're working as a Script Editor, it's all lot more social – you'll have an office life and non-stop meetings with Producers and Writers.

**WHAT YOU NEED TO BE**
Creative, imaginative. Patient, self-disciplined. A big reader. Able to take criticism and, if you're an editor, give it.

**WHAT YOU NEED TO HAVE**
Strong written and communication skills. Good research skills, with an ability to think structurally.

# ARTISTS

If you're lucky enough to have the artistic gene, how about working in one of these? Art and Design, Costume/Wardrobe, Make-up and Hair or Set Crafts.

## THE JOBS

- **Art and Design**: Production Designer, Art Director, Set Designer, Graphic Designer, Model Maker, Props Master/Mistress, Set Decorator/Dresser, Art Department Assistant or Runner.
- **Costume/Wardrobe**: Costume Designer, Costume/Wardrobe Supervisor, Costume Design Assistant, Costume Maker, Wardrobe Assistant, Dresser.
- **Make-up and Hair**: Make-up Designer, Make-up Artist, Special Effect Make-up Artist, Wig Specialist, Make-up & Hair Assistant.
- **Set Crafts**: Construction Manager, Carpenter, Joiner, Set Painter and Decorator, Plasterer, Wood-working Assistant.

## THE LOW-DOWN

How many people work in these jobs? According to the Skillset Employment Census 2002, there are about 3,500 people working in Art and Design, about 1,000 in Costume/Wardrobe, about 1,100 in Make-up and Hair and 500 in Set Crafts. That's across both the TV and film industries.

The majority are freelance. Ninety per cent of Make-Up and Hair professionals, for example, are self-employed. However, there is slightly more stability in Art and Design, where 50 per cent are in jobs rather than freelance. Costume and Make-up are still very female professions – 82 per cent of Wardrobe staff, and 93 per cent of Costume staff are women. Set Crafts, on the other hand, are still dominated by men.

One of the things people love about these jobs is the variety. It's project work, and every job is different, with new challenges. You're unlikely to spend much time in offices – instead, jobs are mostly based in workshops, on set or on location. There can be lots of travelling, with long stints spent on location far from home.

## WHAT YOU NEED TO BE

Artistic *and* practical. Full of ideas and solutions to problems. Good with your hands, and great at teamwork. Flexible. Independent and self-sufficient enough to survive the freelance life.

## WHAT YOU NEED TO HAVE

Training – in art and design, crafts or make-up, depending on what you go into.

# TECHIES

If you've always loved fiddling about with gadgets and gizmos, pressing buttons, tweaking switches, taking things apart and finding out how they work, how about doing something technical in TV?

## THE JOBS

- **Camera**: Lighting Camera Operator, Camera Supervisor, Camera Operator. Camera Assistant, Focus Puller, Grip, Steadicam Operator.
- **Lighting**: Lighting Director, Gaffer, Best Boy, Lighting Assistant.
- **Sound**: Sound Supervisor, Sound Recordist, Boom Operator, Sound Assistant.
- **Studio Operations**: Vision Mixer, Vision Control Engineer, VT Operator.
- **Post-Production**: Editor, Assistant Editor, Dubbing Editor, Dubbing Mixer, Post-Production Supervisor.
- **Broadcast Engineering**: Studio Engineer, Transmission Engineer, Post-production Engineer.

## THE LOW-DOWN

Some of the people in the above jobs might be a bit miffed at being put in the 'technical' section of this book – what many of them do, especially in the more senior jobs, is both highly technical and highly creative. That said, almost everyone in this field works their way up from the bottom and the jobs are much more technical than creative at first.

How big is the technical workforce in the audio-visual industries? Well, according to the Skillset Employment Census 2002, there are about 2,600 people working with cameras, 1,700 working in lighting, 1,300 working in sound, 2,200 in studio operations, 7,200 in post-production and 2,200 in broadcasting engineering.

Nearly all the roles involve being in a studio or set environment, so if it's an office job you yearn for, think again. With the exception of the post-production roles, these are jobs where you will mostly be on your feet and physically active.

The good news is that this is, on the whole, where the demand is – the technical workforce in TV is ageing rapidly and there is a shortage of skilled younger people to fill their boots. The problem, says the industry, is not that there aren't people who want to do these jobs, but that they don't have the right skills to do them. This is partly because there aren't nearly enough training schemes. If you can get on one, you will (in most of these jobs) be in high demand.

**WHAT YOU NEED TO BE**
Technically-minded but also, for many of the camera and post-production jobs, creative too. You need to love fixing problems and coming up with solutions, often under pressure. For some jobs, you also need physical strength.

**WHAT YOU NEED TO HAVE**
Nearly all these jobs require training, although often you can train on the job, in junior roles. Generally speaking, qualifications in maths and physics can help, along with good IT skills and, if you're interested in camera work, experience in photography.

## BUSINESS HEADS

If you've got a good head for business, for buying and selling, for planning, promoting and number crunching, and generally coming up with brilliant ideas and strategies, how about working in the one of these areas.

**THE JOBS**
- **Commissioning**: Commissioning Editor, Development Executive
- **Scheduling**: Scheduler, Planning Assistant.
- **Programme Distribution**: Sales Director, Acquisitions Assistant, Forward Planner or Programme Co-ordinator.
- **Finance**: Production Accountant.
- **Marketing, Sales and Advertising**: Sales Manager, Marketing Manager, Advertising Manager, Sales Representative.
- **Press and Public Relations**: Public Relations Officer.
- **Human Resources**: Human Resources Manager, Personnel Officer.
- **Legal and Rights**: Lawyer, Legal Assistant, Legal Secretary, Rights Assistant.

## THE LOW-DOWN

This is where you'll find jobs aplenty. Many people assume a career in TV means programme-making, but there are thousands of other opportunities out there, and if any of these grab your fancy, they can be easier to get into that the programme-making roles – and just as creative. There are about 27,000 people in these jobs, including all administrative, secretarial and general management jobs in the business, and nearly all of them are employed, rather than freelance, so there's also a lot more stability than in programme-making.

More specifically, there are about 350 people working in Commissioning and 550 working in Scheduling, according to Skillset. For some reason, there are high numbers of women in these roles. What do they do? They plan, commission and schedule broadcasters' schedules, in the short and long term. Commissioning Editors are often former programme-makers – Producers and Editors – who've been promoted. Schedulers may come from sales or marketing backgrounds, or may have started out in Scheduling, maybe as a Planning Assistant.

There are about 550 people working in Programme Distribution, selling and marketing programmes to broadcasters in the UK and abroad. Most people come from a sales and marketing background.

There aren't stats for the number of people working in Finance. They tend to be qualified and part-qualified accountants who have sometimes already worked for other media or accountancy firms. Skilled and experienced Production Accountants are much in demand – now that budgets are so tight, they play a crucial role in managing budgets and making sure programme-makers don't overdo it. There are some specific training opportunities for them – see Chapter 6.

Again, it's not known how many people work in general TV Marketing, Sales and Advertising roles, but the numbers are probably high and the range of jobs broad. Lots of people have worked in marketing, sales and advertising in other industries, and then moved into TV. The same goes for Human Resources, Public Relations and Legal staff. Some PR staff have a journalism background. Lawyers have a legal training, often moving sideways from a private practice.

## WHAT YOU NEED TO BE
A confident communicator and, in some roles, negotiator. Organised, efficient and good at planning. For some roles, an analytical, logical problem solver.

## WHAT YOU NEED TO HAVE
Some of these jobs require specific training, in law or accountancy, for example. Others require a more general, non-TV background, in sales and marketing or public relations. But they *all* require an interest, passion even, for TV and the programming you'll be helping to plan, produce, sell and promote.

# NEWS HOUNDS

Are you always up on the news, scanning the headlines in the papers, tuning in to hourly radio news bulletins? Do you love finding things out, writing things down and talking about current affairs with other people? You may have the makings of a broadcast journalist.

## THE JOBS
News Editor, Producer, Presenter, Correspondent, Reporter, Researcher, Broadcast Assistant.

## THE LOW-DOWN
There are about 10,000 people working in broadcast journalism and sport, according to the Skillset Employment Census 2002, and only about 15 per cent of them are freelance. Jobs are equally divided between men and

'If you're thinking about a career in journalism generally, broadcast journalism offers much better pay and many more jobs than print journalism.'

women. The number of jobs has risen in recent years, thanks to all the new digital and rolling news channels. At the same time, the last decade has seen a big shift towards 'bi-media' production (getting journalists to produce reports for both TV and radio) and 'multi-skilling' (getting journalists to do their own filming, sound, editing, etc.), so some jobs have been consolidated.

The good news is that, if you're thinking about a career in journalism generally, broadcast journalism offers much better pay and many more jobs than print journalism. Jobs on local and national papers are few and far between, whereas there are some major employers in broadcast journalism – *numero uno* being the BBC.

The working environment is as you might guess – bustling. Most offices are big and open plan, and on the whole, people don't have their own desks. Lots of the work is shift work, so the hours can be a nightmare if you're working on an early news programme that needs you to be in work by 5.30am.

One common misconception is that journalists are always out and about, chasing stories. If you're a reporter, you will do a lot of rushing around, but lots of journalists spend much of their time in the office, writing bulletins from news that comes in on the wires (electronic news services), 'packaging' reports up from any available footage, writing scripts and 'links' for presenters, and setting up live interviews.

Some broadcast journalists stay in news rooms for life. Others move into news programmes and documentaries.

### WHAT YOU NEED TO BE
A news junkie. Curious and inquisitive. Persistent and persuasive. Sceptical and (as) objective (as possible). Accurate and able to work under stress. Self-confident. If a reporter or presenter, then something of a performer.

### WHAT YOU NEED TO HAVE
Almost all journalists have degrees, but they don't need to be media-related. They also need a postgraduate training in broadcast journalism. Some succeed without, although employers increasingly demand it. Most people get their first break in radio, working there for a few years before moving into TV.

'You're going to have to polish up your job-hunting skills and learn how to schmooze your way into work.'

# here's what you need to know

So now you know what jobs are out there, how are you going to get one? TV is a competitive business. The trick is to get a sound, well-rounded education, and as much training and work experience as you can. You're also going to have to polish up your job-hunting skills and learn how to schmooze your way into work. Networking is key.

Above all, you're going to have to show enthusiasm and initiative – no one's going to hand you training, or a job, on a plate. One way to show willing is to start working on your own TV projects now, by making your own documentaries, writing your own scripts or pitching your own ideas to producers.

This chapter will tell you how to do all that... and more.

# Insiders' tips on getting in

'Being a runner is the best way to start in television; you get a real feel for what you do and don't like.'

*Ed Suitor, Location Manager*

'I'd say try and get some experience in local radio. It's a very good place to start. Some of the post-graduate courses are very good. I wouldn't recommend a media studies course. You should do something which broadens your mind – a language, history or politics. Anything other than the media.'

*Jon Snow,* Channel 4 News *presenter*

'It is all about experience and making contacts. There's rarely a quick route. It's all about getting a foot in the door, slogging yer bum off and hoping your hard work and suggestions get you noticed.'

*Meera Syal, Actress, Novelist and Screenwriter*

'Unfortunately, the best way into television production is often having a contact, but work experience also leads to good jobs. Make sure you go to a good, reputable company – it doesn't matter if they're small or big.  With the smaller ones, take your CV personally – you put them more on the spot.'

*Michaela Furney, Researcher*

'Bombard people with tapes. Get a friend to video you, pick the best two minutes of footage and send it everywhere. Repeat this process every four months.'

*Claudia Winkleman, Presenter,* Liquid News

'Courses are good for getting motivated. It can be very hard to get going on your own. Start by writing an episode for a soap or a short film. There are loads of short film schemes you can apply for. Or try and get someone at film school to make it. It doesn't matter if what you write at first doesn't get made – it's your calling card to get you other work.'

*Amy Jenkins, Scriptwriter, including* This Life *and* Elephant Juice

'Get a good art training – in drawing, painting, history of art, sculpture. Find an inspiring designer to work for and offer your services to them. Find out where they're working and try to meet them face-to-face.'

*Alison Dominitz, Production Designer*

'Never be scared to send your stuff off to people. You always need a new project on the go because most of them get rejected. Go to as many talks, courses and events as you can and, if you can bear it, try to talk to people. Be friendly and nice. People are often scared of writers – they think you're going to be a freak.'

*Jesse Armstrong, Comedy Writer, including*
Peep Show *and* Smack the Pony

'Be persistent. Write letters, make phone calls and follow them up on a regular basis. If you can get work experience in any department it will enable you to get a foot in the door and hopefully have a look around other departments within the company.'

*Phil Hobden, Head of Presentation, BskyB*

## QUALIFICATIONS, COURSES AND TRAINING

There are hundreds of media-related qualifications and courses out there. Sadly, there aren't nearly as many jobs, so if you are going to train – especially if you're paying for it yourself – it's important you do the right kind of training on the right kind of course.

Finding out about training opportunities and courses is in itself a full-time job. There are so many professional bodies, industry organisations and course providers, it can all feel a bit overwhelming. Let's kick off with some of the basic, frequently asked questions.

### FAQS

### What should I study?

It depends what you want to do in TV and whether you're planning to go to university. Either way, it's advisable to get some core GCSEs – English, maths, a science, a foreign language and ICT. If you are planning to go to university, you'll need to get enough A-levels to do the course you want, so think about what kind of degree you're interested in and which A-levels are recommended for those courses.

Media studies is not a must, at GCSE or A-level, but try and find out if the university courses you're interested in prefer you to have them. As a general rule, you should do what you're most interested in, what you love – it doesn't matter if it's not directly related to TV or the media.

For some more practical jobs, you might be in a stronger position later if you take General National Vocational Qualifications (GNVQs) and Advanced GNVQs, a.k.a Vocational A-levels. These give you more practical, vocational experience. However, if you are planning to go on to university, make sure the courses you're interested in rate Advanced GNVQs as highly as A-levels.

If you're thinking of doing something technical, it's probably a good idea to get some maths, science and technology qualifications. Even if you're not planning to do any further education, GCSEs in these subjects may give you a head start when you're applying for training or jobs.

### Do I need a degree?

You definitely don't *have* to have an honours degree to work in TV. Instead, you could get NVQs or SVQs, a Modern Apprenticeship or a

Foundation Degree, relevant to whatever it is you want to do. There are also some very good new entrant training schemes open to over-18s (see page 73).

Be warned – you'll have to fight to get a place and many applicants are often over-qualified. Although you don't need a degree to get on a training scheme, or to get a job, lots of people applying have them. Fifty-six per cent of people working freelance in broadcast TV have a degree, and 26 per cent also have a postgraduate qualification, according to the Skillset Freelance Survey.

There are some areas where a degree is an absolute must – in broadcast journalism, for example, or in art and design, where an art degree is a basic requirement. Likewise for some highly specialist, technical jobs, a science degree is the norm. This is something you will need to really research. See Chapter 8 for some ideas on where to find out more.

Bear in mind that, apart from qualifications, there are other important things a degree can give you. Further education develops your mind, teaching you to think more critically and analytically. It also gives you time – time to think about what you want to do in life, time to work on student radio, TV or films, time to get some work experience in the holidays, and time to develop your social skills. Finding a job and working in the media are all about confidence and getting on with people – life skills not everyone has when they're fresh out of school.

'Apart from qualifications, there are other important things a degree can give you. Further education develops your mind, teaching you to think more critically and analytically.'

## Should I do a media degree?

Not necessarily. Nearly 23 per cent of freelancers working in broadcast TV have a media-related degree, according to the Skillset Freelance Survey. More – 33 per cent – have other sorts of degrees.

'A degree in media studies is not essential,' says PACT, the Producers Alliance for Cinema and Television. 'The best media studies degrees with a vocational bias are generally at postgraduate level... Another degree such as law, business or a language might offer you more flexibility.' In other words, you might be better off doing a non-vocational first degree, followed by a vocational postgraduate degree.

There are some good media-related degrees out there, but many employers are sceptical. As Jonnie Turpie from Maverick TV says, 'Good qualifications are always impressive. However, "good" in the media does not always refer to a media degree, it could be English, Chemistry or even Anthropology.' There is a growing feeling that, because so many people have them, they are not worth as much as other degrees. Some see them as an easy ride – much easier to get than a language or science degree. Also, programme makers want to recruit people who know something about life and the world *beyond* the media. They want people with knowledge that can be used to make interesting programmes.

# Career Training and Qualifications

Where to find advice:

- **IDEAS**FACTORY has an online advice service, in association with learndirect, answering questions about training and careers in the creative industries. See the Training & Courses section at www.ideasfactory.com.
- **skillsformedia** is a brilliant source of advice for new entrants, students and people already working in TV. It is run by Skillset, BECTU and the government's national learning advice line, learndirect. Its website has information on careers, qualifications and training. It also offers one-to-one advice sessions over the phone with specialist media advisers. See www.skillsformedia.com or call on 08080 300 900.
- Anyone aged up to 19 in England can get advice on qualifications and courses from **Connexions**, the support service for young people. See the Connexions website (www.connexions.gov.uk) for information and to find out where your local Connexions Service is.

If you are thinking about a media-related degree, it needs to be one with a strong practical, vocational element to it. It's important to understand the difference between media studies degrees and media degrees. Media studies degrees tend to be purely academic and theoretical, analysing the role of the media in society, whereas media or media production degrees are more practical, teaching professional skills. See Choosing the Right Course on page 71.

### Can't I just learn on the job?

Yes. You could get a job as a runner or a trainee, or a secretarial role, and work your way up. The danger is that you hit a barrier later on, because you don't have qualifications. The answer is to learn and train on the job.

This is easier said than done. In recent years, because so much of TV is freelance, on-the-job training has been hard to get – people have picked up whatever skills they can, doing the odd course here or there. That is beginning to change – vocational qualifications, and in particular the Skillset Professional Qualifications, which are NVQs and SVQs you get while working, are taken increasingly seriously (see below). In future, they may be a must for anyone wanting to move up the career ladder.

### VOCATIONAL QUALIFICATIONS

There are lots of media-related vocational qualifications you can get before you start working – GNVQs, NVQs or SVQs, City & Guilds, BTECs and Foundation Degrees. To find out about all of them, speak to a careers adviser. You can also look up courses on the UCAS database (www.ucas.ac.uk) or in the reference book, *British Vocational Qualifications* (Kogan Page).

For information on NVQs, speak to learndirect (Tel: 0800 585 505) or see the NVQ website (www.dfes.gov.uk/nvq). For information on City & Guild courses, see www.city-and-guilds.co.uk. For information on BTECs, see www.edexcel.org.uk. For information on Foundation Degrees, see www.foundationdegree.org.uk.

### Skillset Professional Qualifications

Once you're working, you can get Skillset Professional Qualifications (NVQs) – the 'gold standards' for the television, film, video and multimedia industry. These can't be done at college, only once you're working in the industry. Nor are they done by sitting exams – you are assessed on the job by an approved assessor from one of Skillset's nine Assessment Centres around the country.

Each NVQ, which come at various levels, relates to a specific job, and is proof for future employers that you can do that job. More than 3,000 people already have them. The hope is that by making the Skillset NVQs a must-have, employers will have a much easier time recruiting people – they'll know exactly what candidates are capable of.

To find out more about the Skillset NVQs, see the Skillset website (www.skillset.org). Check out the Route Maps there – these list the Skillset NVQs and show you exactly what experience you need to achieve them. For advice on doing the Skillset NVQs, you could also speak to a skillformedia adviser (Tel: 08080 300 900).

### Modern Apprenticeships
Modern Apprenticeships are for over-16s who want to train while working. There are two levels of Modern Apprenticeships: Foundation (FMA) and Advanced (AMA). Both lead to NVQs and other qualifications. There is a Modern Apprenticeship in broadcast, film, video and multimedia, which can be done while working in, for example, Post-production, Camera Assistance or Make-up.

To find out more, call 08000 150 600 or see the Modern Apprenticeships website (www.realworkrealpay.info). For information on Modern Apprenticeships in Scotland see www.modernapprenticeships.com/, in Wales www.beskilled.net and www.elwa.org.uk/, and in Northern Ireland www.delni.gov.uk.

### COURSES
Deciding whether or not to study a media, TV or journalism course full-time is a big decision, especially if it's a postgraduate course you're paying for.

If you do decide to do a course, it's important to make sure you choose the right one – one that will teach you genuinely useful skills, using real industry-standard equipment.

Unfortunately, there is no single, gold standard accreditation system for media courses, although Skillset is working on one. That means you're going to have do a lot of research. (For accredited broadcast journalism courses, see below.)

Remember, higher education is big business these days – course providers and their syllabuses are selling you their services, so make sure you ask some hard-hitting questions before applying.

## Choosing the Right Course

- Read the prospectus carefully and check out the *Media Courses UK* guide (see below) for the low-down on the course. How much of it is practical? Study the syllabus. One sign of a good vocational course is that the syllabus is modelled around, or similar to, the Skillset Professional Qualifications – a good indication that it's gearing up students to do real jobs in the real world. To read more about the Skillset standards, see www.skillset.org.
- What kind of equipment do they have? Visit the facilities – does it look as if they have enough equipment for the number of students they take?
- Ask what industry links the course has. Do they help students find work experience? Where have previous students done placements?
- Ask what's happened to previous students – how many of them found jobs and doing what? Try and find former students to speak to about what was useful and useless about the course.
- What experience do the tutors have? Have any of them worked in TV in recent years – if so, what on? Good courses often have tutors who've worked in the business, or are still working in it.

**❝**   MY EXPERIENCE – SOUND
RECORDIST AND BOOM OPERATOR

'The trouble with lots of courses is that they teach skills across the industry, so students leave with a very superficial knowledge. People come out of college and are really disappointed to find that out. Also, there's an over-supply of courses. I think the colleges are to blame – they take the money and promise careers that just aren't out there.'

*Becky, Sound Recordist and Boom Operator*

**❞**

## Finding out about Courses

These are good places for doing research into the right course for you:

- *Media Courses UK* by Lavinia Orton (BFI) – an annual and comprehensive guide to media courses, based on Skillset and the British Film Institute's database of courses. Alternatively, you can search the Skillset/BFI database online for free, by place, institution or keyword, at www.bfi.org.uk/education/courses/mediacourses.
- **IDEAS**FACTORY (www.channel4.com/ideasfactory) has a course-finding service in its Training and Courses section. You can search or browse through hundreds of options.
- Good broadcast journalism courses, undergraduate and postgraduate, are accredited by the Broadcast Journalism Training Council – there is a list on its website (www.bjtc.org.uk). To find out which ones the industry rates, see the list of courses that the BBC sends the trainees on its News Sponsorship Scheme on, at the BBC Jobs website (www.bbc.co.uk/jobs/bbctrainees).

## Getting Funding

Tuition fees for courses can be expensive, not to mention your living costs while doing them. If it's your first, undergraduate degree you'll get some state funding, but for postgraduate or short courses there are, sadly, no big training funds. Some courses have a few grants and bursaries, but it will probably be up to you to raise the cash. Here's where you can get some help:

- Career Development Loans – these are for between £300 and £8,000, for up to two years of training. Barclays, The Co-operative or The Royal Bank of Scotland lend the money, but the Department for Education and Skills pays the interest. To find out more, call the LearnDirect advice line (Tel: 0800 585 505) or see www.lifelonglearning.co.uk/cdl.
- For more information on getting financial help for training, get the Department for Education and Skills booklet *Money to Learn* by calling 0845 602 2260 or at www.lifelonglearning.co.uk/moneytolearn.
- There are charities you could write to, to ask for sponsorship. Try your local library for *The Directory of Grant Making Trusts*, the *Charities Digest* and the *Hollis Sponsorship and Donations Yearbook*.

## NEW ENTRANT TRAINING SCHEMES

If you can get onto one of these, you'll be well on your way. New entrant schemes are formal training schemes, often in specialist areas – camera, sound, wardrobe, etc – although there are some general production ones too. Trainees usually do short, highly focused vocational courses, but spend most of their time working on real programmes or projects, in real jobs.

The bad news is that, according to skillsformedia, there are fewer than 100 places on all the training schemes put together – and thousands apply. 'Applicants that have done their homework on the industry, are keen and enthusiastic, know their area of specialisation, and perhaps already have some work experience, will stand a far better chance,' it advises.

The main schemes are outlined below.

### BBC

The BBC used to be the main training provider in the UK. It still trains about 350 new recruits a year, but training schemes now tend to be offered when the need arises, rather than year in, year out. All schemes are advertised in the press and on the BBC's Job pages online (www.bbc.co.uk/jobs), so keep an eye out. These are some of the schemes they have been running recently:

- *Programme Making* – Two-year general programme-making trainee schemes, including training and placements, in London, Glasgow, Cardiff and Belfast. Also, Entertainment Trainee schemes in Glasgow, Cardiff and Belfast. At the end, trainees are qualified to be Assistant Producers.
- *Engineering and Technology*:
  - *Pre-university* – paid, gap year training, with the opportunity to be sponsored through university and do vacation work.

## ONLINE COURSES

There is a lot you can learn about TV, without even having to leave your bedroom:

- The Global Film School (www.globalfilmschool.com) is an online university dedicated to filmmaking, using the most highly respected educators and training in the field. It is run by the UCLA School of Theatre, Film and Television, in conjunction with the UK National Film & Television School and their Australian counterpart.
- The BBC Training & Development website (*www.bbctraining.co.uk*) has some of the courses it uses to train staff online – and for free. There are dozens to choose from, for example Pre-production, Post-production and Introduction to Video Production.
- **IDEAS**FACTORY has teamed up with Vision2Learn to offer **IDEAS**FACTORY visitors free online courses, mostly in career development skills. See the Training & Courses section on the **IDEAS**FACTORY website (www.channel4.com/ideas factory).

- *Vacation Training* – paid training during university holidays.
- *Industrial Placements* – year-long, paid placements.
- *Graduate Trainees* – schemes in Broadcast Engineering (London, Cardiff, Glasgow, Edinburgh, Aberdeen, Inverness, Belfast and Tunbridge Wells); in BBC Technology (West London); and in Research and Development (Surrey).

● *Resources and Operations.* Two-year scheme with BBC Resources Ltd, a commercial subsidiary of the BBC providing production facilities and design services. Trainees work in one of these:
  - *Studios* - specialising in either cameras, sound or lighting, with a view to employment as a Camera Operator, Sound Assistant or Vision Operator.
  - *Outside Broadcasts* – specialising in 'OBs', with a view to becoming a Technical Operator.
  - *Post-production* – specialising in viewing, transferring, transmitting and digitising video and audio material, with a view to working as an Editor.

● *News and Journalism:*
  - *News Sponsorship Scheme* – sponsors trainees to do a postgraduate diploma in broadcast journalism, on one of a choice of nationwide courses, plus five weeks of work experience.
  - *Broadcast Journalist Trainee Scheme* – one-year local and regional news training scheme, with courses in Bristol followed by placements in one of 14 newsrooms in the UK.
  - *Current Affairs Trainee Researchers* – one-year scheme working in TV, Radio or Online.
  - *Broadcast Technology Assistant Trainees* – two-and-a-half year scheme working for News Resources, which provides technical support in newsrooms and studios.

● *Vision Design* – A one-year scheme for young designers, with training and on-the-job experience. Selection is done through an annual Design Competition followed by assessed work attachments. In 2003, it took 13 trainees in art direction, costume design, make-up and new media design, plus two in broadcast graphics and interactive design. For more information, see the Vision Design website at www.bbc.co.uk/designvision.

For more information on all these schemes, see the Trainees section on the BBC Jobs website (www.bbc.co.uk/jobs/bbctrainees) or call 0870 333 1330.

## CHANNEL 4 TRAINING INITIATIVES

Channel 4 Television is one of the leading broadcasters in supporting and developing new talent in the audio-visual industries. Channel 4 places a strong emphasis on nurturing talent. The following initiatives exemplify the new-entrant training schemes. It's important to note that these schemes do not necessarily run every year but are evolved to respond effectively to the current skills shortages or key areas of under-representation.

1. *Junior Researcher Scheme.* This year-long scheme is designed to encourage culturally diverse talent and aims to attract new entrants from ethnic minority backgrounds into the industry.

   Entering at Runner level, the scheme provides an all-round grounding in the basics of television production and equips participants with skills to carry out the role of a junior researcher. Twelve participants take part in the programme, each placed with an independent production company.  Participants receive additional training at Channel 4.

2. *Researcher Training Initiative.* A one-year scheme, this too is aimed at culturally diverse talent with some production research experience.

   The six participants are based within independent production companies and each works as a researcher in production and development research. The aim is to develop the necessary skills and knowledge to allow them to move up the production ladder to Assistant Producer level.

3. *Other Channel 4-related schemes.* Channel 4 also supports and sponsors a wide variety of training schemes with a number of independent organisations. These range from music promo production initiatives to animation schemes, from acting classes to trainee drama director opportunities, plus regional support schemes and support for disabled and hearing impaired people in the industry. Details of most of these schemes can be found on **IDEAS**FACTORY (www.channel4.com/ideasfactory).

For more information on these and other training initiatives, see the Channel 4 website www.channel4.com/4careers/faq.html. Also keep an eye out for adverts in the ethnic minority and national press, for example in Monday's *Media Guardian*).

## FT2

FT2 – Film and Television Freelance Training – is another leading training provider for new entrants. It is for over-18s who want to work freelance in the construction, production and technical areas of film and TV. Trainees do placements, short courses and Skillset NVQs. FT2 runs three different schemes:

1. *New Entrant Technical Training Programme.* A two-year apprenticeship scheme, with 10–15 places a year, for newcomers wanting to become:
   - Art Department Assistant
   - Assistant Editor
   - Camera Assistant/Clapper Loader
   - Grips
   - Make-up/Hair Assistant
   - Production Assistant
   - Props Assistant
   - Sound Assistant
   - Wardrobe Assistant
   - Assistant Location Manager.
2. *Set Crafts Apprenticeship Training Scheme.* A two-year apprenticeship, with four to six places a year, for new entrants wanting to work in set construction, in:
   - Carpentry and Joinery
   - Fibrous Plastering
   - Set Painting and Decorating.
3. *Independent Companies Researcher Training Scheme.* Eighteen month scheme for six to eight people, aged 20 plus, wanting to work as researchers in factual programming. Trainees are placed with several of the 14 independent production companies that sponsor the scheme.

For more information on FT2 schemes and advice on how to apply, see the FT2 website or send an A4 S.A.E to FT2, Fourth Floor, Warwick House, 9 Warwick Street, London W1B 5LY, Website: www.ft2.org.uk.

## Cyfle

Cyfle is the leading vocational training provider for the television, film and interactive media industries in Wales. It has a full-time, one-year training scheme, with 18 places. Trainees are placed in production and facility companies in Wales. They also do short courses and are assessed for Skillset NVQs.

Contact: Cyfle, Gronant, Penrallt Isaf, Caernarfon, Gwynedd LL55 1NS, Tel: 01286 671 000, Email: cyfle@cyfle.co.uk, Website: www.cyfle.co.uk.

## Scottish Screen

An 18-month scheme in which trainees specialise in technical, craft, design and production areas, doing short courses, Skillset Professional Qualifications and placements in television production, commercials and feature films. In 2002, it took eight trainees, one in each of the following areas:

- Camera
- Production Factual
- Production Drama
- Assistant Directing
- Make-up
- Post-production
- Art Department
- Locations.

Contact: Scottish Screen, 2nd Floor, 249 West George St, Glasgow G2 4QE, Tel: 0141 302 1700, Email: nets@scottishscreen.com, Website: www.scottishscreen.com.

## Northern Ireland Film and Television Commission

The Northern Ireland Film and Television Commission, an agency for the development of the film and TV industry in Northern Ireland, runs the FOCUS programme, which provides work placements on local productions for trainees who want to learn new skills and develop existing ones. Interviews take place twice a year, although at the time of going to press, the scheme was 'on hold until further notice'.

Contact: Northern Ireland Film and Television Commission, 3rd Floor, Alfred House, 21 Alfred St, Belfast BT2 8ED, Tel: 028 9023 2444, Email: info@niftc.co.uk, Website: www.niftc.co.uk.

## Michael Samuelson Lighting

A highly-rated scheme for new entrants wanting to become lighting professionals. It takes up to ten trainees a year and places them throughout the television and film industry for a year. Applicants need basic electrical qualifications.

Contact: Michael Samuelson Lighting, Pinewood Studios, Iver Heath, Buckinghamshire SL0 0NH, Tel: 020 8795 7020.

### Animator in Residence (AIR) Scheme

Channel 4 and the British Film Institute run this scheme, offering four trainees the chance to do three-month residencies at the National Museum of Photography, Film & Television (www.nmpft.org.uk). Applicants need to be recent graduates (within the last five years) with an original idea for a three-minute animated film. At the end, they submit a proposal to Channel 4.

Contact: The AIR Scheme Administrator, BFI/NT, South Bank, Waterloo, London SE1 8XT, Tel: 020 7815 1376, Website: www.a-i-r.info.

### Sky Finance

A graduate programme for the finance department. Applicants need a 2.1 degree in business or finance. For more information, see www1.sky.com/jobs.

### Assistant Production Accountant Training Scheme

Scheme run by the Production Guild for new entrants wanting to become Assistant Production Accountants. It takes six trainees for 12 months. It does short courses, leading to the Skillset NVQ in Production Accounting Level 3, and also works on big productions.

Contact: The Production Guild, Pinewood Studios, Pinewood Road, Iver Heath, Buckinghamshire SL0 ONH, Tel: 01753 651 767, Email: info@productionguild.com. Website: www.productionguild.com.

### Other Training Opportunities for New Entrants

There are other new entrant training schemes around the country. These are the organisations you should contact to ask for more information:

- **Media Training North West**, Tel: 0161 244 4637, Email: info@mtnw.co.uk, Website: www.mtnw.co.uk.
- **Northern Film & Media**, Tel: 0191 269 9200, Website: www.northernmedia.org.
- **Screen Yorkshire**, Tel: 0113 294 4410, Website: www.ysc.co.uk.
- **Screen West Midlands**, Tel: 0121 766 1470, Email: info@screenwm.co.uk, Website: www.screenwm.co.uk.
- **South West Screen**, Tel: 0117 952 9977, Email: info@swscreen.co.uk, Website: www.swscreen.co.uk.

- **Screen East**, Tel: 0845 601 5670, Email: info@screeneast.co.uk, Website: www.screeneast.co.uk.
- **EM Media**, Tel: 0115 934 9090, Email: info@em-media.org.uk, Website: www.em-media.org/.

## The SIF Network

If you're lucky enough to get onto one of the above schemes, you may be able to get your name put onto the SIF Trainee Network, run by the Skills Investment Fund (SIF), a database of trainee-level production staff that companies can browse through when they are looking to hire people. For more information, see www.sifnetwork.org.

## DOING IT FOR YOURSELF

If you're itching to get going in TV, don't wait around waiting to be discovered. It could take years, or worse still, it might never happen. Start making TV now.

By making your own programmes, writing your own scripts or pitching your own ideas to other people, you'll put yourself on a steep learning curve – and you'll also have something to brag about when you apply for courses or jobs.

These are just some of the things you could do for yourself:

- write a short script
- film a short
- make a documentary
- make 30 seconds of animation
- pitch an idea
- do a radio report
- get involved in community TV.

## WRITE A SHORT SCRIPT

Write a 10-minute script or, if you're an aspiring comic, a few short sketches. If that feels like going in at the deep end, build up to it by starting a writer's notebook – jot down ideas for characters, plot and dialogue. If you end up with a script you really rate, maybe you could get an aspiring film maker to shoot it, or make it yourself. For competitions you could send your work to, see page 84.

### Online advice for Scriptwriters
- **Raindance** (www.raindance.co.uk) – see the Indie Tips section for advice on scriptwriting and filming.
- **Trigger St** (www.triggerstreet.com) – a web-based film community, set up by Kevin Spacey, where scriptwriters and filmmakers can upload their script or film and get it reviewed.
- **Scriptfactory** (www.scriptfactory.co.uk) – one of Europe's leading development organisations working to support screenwriters by finding and developing new talent.

### Insider's Tip
'Write what makes you laugh. Read dialogue aloud. Live, eat and breathe your characters, talk about them, know everything about them. Get a script, look at it and see how it is presented. A half hour comedy script for the BBC is about 45 pages long. If you're writing alone, get someone you trust to read it and give you his or her opinion.'

*Jan Etherington, Comedy Writer*

## ON THE WEB

- **Filmmaking.net** (www.filmmaking.net) – a site for new and independent film-makers, with a brilliant FAQ section.
- **Exposure** (www.exposure.co.uk) – everything you need to know to shoot your film, including an 'Eejit's Guide to Film-Making'.
- **Reach for the Sky** (http://rfts.sky.com) – Sky's initiative to support and inspire young people to achieve their potential. Contains advice and information about jobs in Film and TV.
- **Making Movies** (www.channel4.com/film/makingmovies/index.html)
- **Shooting People** (www.shootingpeople.org)

# DIY TV

Just because the BBC won't show your first offering in a prime-time slot, it doesn't mean the rest of us should be deprived. Take inspiration from two comedy writers, Jane Bussmann and David Quantick, who put their sitcom, The Junkies, a show about 'the stupidest heroin addicts in the world', online.

'The Junkies is a DIY sitcom,' they say. 'There was no commissioning editor, no broadcaster and no top comedy company involved, just a team of people who worked for no money and very little food. A traditional half-hour comedy costs £200,000 in crew, locations, studio, cast, script and someone to go to Starbucks every five minutes for no reason. "The Junkies" was completed on a budget of £3,500. Anything we didn't have to pay for we didn't use or we borrowed or begged for free.'

*(Find out more, and watch* The Junkies, at www.junkiestv.com and www.thejunkies.com.)

## FILM A SHORT

If you're interested in directing, or being a camera or sound person, the fastest way to learn is to make a short film. There has never been a better time to do it. Video and digital cameras have made low-budget, broadcast-quality filmmaking a reality – and once you've finished shooting, you can easily edit your film on a PC or Mac, and then showcase it on the Internet.

Get friends to help you out, or find volunteers online. If you can, get hold of a copy of *The Guerilla Film Maker's Handbook* (Continuum) by Genevieve and Chris Jones, a bible for low-budget film makers. When it's finished, send it to a film festival or enter it for a competition. See page 83.

## ON THE WEB

- **Documentary Filmmakers Group** (www.dfglondon.com) – articles, forums and links to online documentaries.
- **Docos.com** (www.docos.com) – industry news and information.

## MAKE A DOCUMENTARY

If factual programming is where you want to be, make your own short documentary. Borrow a video or digital camera and edit it on a PC or Mac. Before you start recording acres of footage, do some research – the key to a good documentary. If you need inspiration, look through the national and local papers for story ideas, or try and think of a subject close to your heart or home. After you've recorded it, you'll need to write a script.

## MAKE 30 SECONDS OF ANIMATION

'If people are really interested in getting into this industry, then the best thing they can do is start doing it at home,' said Matt Aitken, a 3D Animator, interviewed on **IDEAS**FACTORY (www.ideasfactory.com). 'If you've got access to a computer, there's free animation software you can get your hands on (try www.tucows.net) – play around and see what you can do. Work towards producing a showreel, which is just getting some work onto a tape. It could be 30 seconds worth of work... It should be concise and really show your talent.' For festivals and competitions you could send your work too, see page 84.

## PITCH AN IDEA

Whatever your age or experience, there is nothing to stop you pitching ideas to broadcasters and independent production companies – as long as you do it like a professional. Everyone in TV is always hungry for original stories and new ideas for series and formats.

To find out how to go about it, see 4Producers (www.channel4.com/4producers), Channel 4's site for Producers, and the BBC equivalent, BBC Commissioning (www.bbc.co.uk/commissioning), which has a section on how members of the public can pitch ideas. If you're already working in the industry, maybe you're a Runner or other first-jobber, this site will be useful for you too.

## DO A RADIO REPORT

If you want to be a journalist, you're much more likely to get a story on the radio than on TV, so how about doing a report for a local radio station? Listen closely to the station for a few days to see what style their news bulletins use and whether there are any news programmes, such as a breakfast news show, you could send your story to.

You will need an original idea, a story that they haven't done yet. Find out the name of the Editor or a Producer and call or email in your story idea –

# Insider tip

Jess Search is Editor of Channel 4 Independent Film & Video, a department charged with commissioning programmes 'outside the mainstream'. Here are the top five things she says you should consider when pitching ideas:

1. Can you describe the idea in a couple of lines? If not, you probably haven't nailed it yet.
2. What's the film called? A clear title helps to directly communicate what the idea will deliver and to demonstrate how it will stand out in the TV listings.
3. What's fresh and contemporary about the idea? Channel 4 looks for ideas that feel relevant and timely.
4. Have you thought about how the film can start, develop and resolve over an hour? Remember, a Channel 4 film has four, roughly 12-minute parts. How, will the story move on and change over each one?
5. Have you been clear about the style and tone? Is it driven by observation, by interview, or by the director's own commentary? It helps to name recent films that can give guidance to your approach.

(*Source: **IDEAS**FACTORY at www.channel4.com/ideasfactory*)

if they like it, they might commission you. Alternatively, record it yourself and send the tape in. You may get knocked back, but any worthwhile Editor or Producer should welcome local stories from local people.

## GET INVOLVED IN COMMUNITY TV

There are dozens of community TV projects, and even quite a few community TV channels, out there – good places to volunteer to work on TV projects or to get your own programme shown. To find ones near you, see the Community Media Association (www.commedia.org.uk).

## ON THE WEB

- BBC Radio (www.bbc.co.uk/whereilive) – details of your local BBC station.
- Community Media Association (www.commedia.org.uk) – listings for local community radio stations.
- Independent Radio News (www.irn.co.uk) – the official website of the independent radio news service.
- BBC Training (www.bbctraining.co.uk) – free online courses on recording and editing radio, plus a BBC style guide with dos and don'ts.

**ON THE WEB**

For international festival listings, check out:

- **International Television Festivals Association** – international festival listings – www.televisionfestivals.com.
- **Filmfestivals.com** (www.filmfestivals.com) has international festival listings, plus news of what's showing where.

You could also pitch ideas or send programmes you've made to the Community Channel (www.communitychannel.org), which goes out on Sky Digital 684, Telewest 233 and Freeview 46.

## OPPORTUNITIES

Once you've started making TV programmes and short films, or coming up with ideas for them, there are masses of ways to get them – and you – noticed.

### FESTIVALS

- **Edinburgh International TV Festival** – www.geitf.co.uk
- **Sheffield International Documentary Festival** – www.sidf.co.uk
- **Bradford Film Festival** – www.bradfordfilmfestival.org.uk
- **London Film Festival** – www.rlff.com
- Birmingham Film & TV Festival – www.film-tv-festival.org.uk
- **Cardiff Screen Festival** – www.iffw.co.uk
- **Bradford Animation Festival** – www.baf.org.uk
- **Bristol International Animation Festival** – www.animated-encounters.org.uk

### COMPETITIONS, SCHEMES AND AWARDS

#### General

- **BBC Talent** – annual competition offering opportunities to make programmes or contracts to work on them; categories vary from year to year, but in the past they've been for new sports reporters, factual programme presenters, weather presenters, pop video directors, stand-up comedians and sketch writers, animators and drama directors. See www.bbc.co.uk/talent.
- **IDEAS**FACTORY Live – This is the offline incarnation of **IDEAS**FACTORY. It usually takes the form of a series of on-the-ground events, which evolves into a competition. **IDEAS**FACTORY Screenplay is an example of a recent competition: 200 would-be screenwriters in the West Midlands were taken through a series of workshops and masterclasses with

renowned film practitioners, culminating in the production and broadcast on Channel 4 primetime of four of their short films. These competitions can be followed closely online at www.channel4.com/ideasfactory.

- **TVYP** – Television and Young People, an annual scheme giving 18–21-year olds the chance to attend five days of masterclasses and workshops at the Edinburgh International Television Festival. TVYP also runs various one-off workshops, competitions and work placement schemes round the UK. See www.geitf.co.uk/tvyp.

- **RTS Student Television Awards** – The Royal Television Society's regional and national awards for student animation, factual and non-factual programmes. See the RTS website (www.rts.org.uk) or contact The Events Department, RTS, Holborn Hall, 100 Gray's Inn Rd, London WC1X 8AL.

- **Vision Design Competition** – part of the selection process for BBC design trainee posts (see Training, page 74). See www.bbc.co.uk/designvision/competition.

- **Guardian Student Media Awards** – annual competition, including a category for Student Reporter, in association with Sky News. See www.mediaguardian.co.uk.

## For writers

- **TAPS** – Television Arts Performance Showcase competition for ten-minute scripts. See www.tvarts.demon.co.uk.

- **Sir Peter Ustinov Award** – international competition for new writers, under 30, with one-off TV drama scripts for 'a family audience', run by the International Council, which holds the Emmy Awards. See www.iemmys.tv.

- **Tony Doyle New Writers Bursary** – annual competition for new drama writers run by BBC Northern Ireland. See www.bbc.co.uk/northernireland/drama.

- **PAWS Drama Fund** – Public Awareness of Science scheme to encourage new writers to develop proposals for dramas about science or engineering/technology; an £1,000 development award. See www.pawsdrama.com.

## For documentary makers

- **Channel 4's Sheffield Pitch Prize** – Channel 4 documentary competition. Shortlisted candidates pitch their idea to commissioners at the Sheffield International Documentary Festival and the winner gets £30,000 to make

## ON THE WEB

For up-to-the-minute news on competitions and opportunities, see:

- **IDEAS**FACTORY – the 'opportunities' page in the Film & TV section – at www.channel4.com/ideas factory.
- BBC Writersroom – the 'opportunity' section – at www.bbc.co.uk/writersroom.

it. See www.channel4.com/4producers.
- **Alt TV, Other Sides and Outside** – Channel 4 strands showcasing new talent, all commissioned by the channel's Independent Film & Video department. For more information, see the commissioning section of 4 Producers (www.channel4.com/4producers) or write to: Independent Film & Video, Channel 4, 124 Horseferry Rd, London SW1P 2TX.
- **This Scotland** – Scottish Screen, Scottish TV and Grampian TV scheme, offering the opportunity to make 12 half-hour documentaries, to both experienced or novice doc-makers. See www.scottishtv.co.uk/factual or call Scottish TV (0141 300 3000) or Scottish Screen (0141 302 1700).

### For short films
- **Digital Shorts** and **New Cinema Fund/FilmFour Lab** – various UK Film Council short film schemes. See www.filmcouncil.org.uk/shorts.
- **First Light** – UK Film Council funding and awards for short films by young film makers (7–18 years). See www.firstlightmovies.com.
- **Nike Young Directors Awards** – Nike-sponsored short films competition, in association with Britshorts, a showcase site for UK and European shorts. See www.britshorts.com/nike.

### For animation
- **Mesh** – Channel 4's digital and interactive animation competition. See www.channel4.com/mesh.
- **Animate!** – Channel 4 and the Arts Council of England funding scheme for experimental animation. See www.animateonline.org.

## GETTING A JOB

### JOB-HUNTING
You'll have to fight to get a job in TV. There are hundreds of other people out there looking for their first break, and hundreds more who already have experience and are hunting around for their next contract or job.

'You'll have to be plucky and persistent. You'll need to do a lot of research and learn how to sell yourself. That means having a solid CV and good interviewing and networking skills.'

Here's the good news. There are more jobs than in other areas of the audio-visual industries – and more jobs get advertised. In film, the jobs market is almost entirely word-of-mouth. In TV, broadcasters often have equal opportunity policies that oblige them to advertise jobs when they come up.

Not surprisingly, there are more ads for jobs where there are skills shortages, and jobs in administrative and commercial areas. You won't see many for runners or researchers. If that's what you want to be – and if you want to work for an independent production company, where contacts and word-of-mouth are how many people get work – you'll have to be plucky and persistent. You'll need to do a lot of research and learn how to sell yourself. That means having a solid CV and good interviewing and networking skills.

### HOW THEY GOT THERE
When Skillset did a survey of freelances working in the UK's audio-visual industries, it asked them how they got their first break in the industry. Here's what they said:

|                        | Broadcast TV (%) | Independent Production (%) |
|------------------------|------------------|----------------------------|
| Responded to job ad    | 24               | 19                         |
| Approached by employer | 13               | 14                         |
| Contacted employer     | 32               | 31                         |
| Agency                 | 4                | 3                          |
| Friend or relative     | 21               | 27                         |
| Word of mouth          | 10               | 11                         |

|                        | Corporate (%) | Animation (%) |
|------------------------|---------------|---------------|
| Responded to job ad    | 26            | 12            |
| Approached by employer | 12            | 6             |
| Contacted employer     | 33            | 53            |
| Agency                 | 3             | 12            |
| Friend or relative     | 22            | 24            |
| Word of mouth          | 11            | 6             |

*(Source: Skillset Freelance Survey, January 2002)*

## WHERE TO LOOK

The trade press and the Internet are your best bets when job-hunting, but check out local papers too for opportunities with regional companies. The national press is less likely to have ads for entry-level jobs, but keep an eye on Monday's Media section in the *Guardian* and its website (www.jobs.guardian.co.uk) – that's where major training schemes and opportunities get advertised, and there is also the occasional small ad for an entry-level job.

## Weather watchers

Fancy a career as a weather presenter? Check out the BBC's Weatherwise centre (www.bbc.co.uk/weather/ weatherwise), the Royal Meteorological Society (www.royal-met-soc.org.uk) and the Met Office (www.metoffice.com). The BBC also has a Weather Watcher scheme for wannabe weathermen (www.bbc.co.uk/talent).

## TRADE PRESS

- *Broadcast* (www.broadcastnow.co.uk)
- *AV – Audio Visual Magazine* (www.avmag.co.uk)
- *Screen International* (www.screendaily.com)
- *Campaign* (www.brandrepublic.com)
- *Media Week* (www.mediaweek.co.uk)
- *Press Gazette* (www.pressgazette.co.uk)
- *Ariel* – the BBC's in-house magazine, available by subscription (Tel: 01709 364 721) or you can pick it up for free in any BBC building reception
- *The Stage* (www.thestage.co.uk)
- *Televisual*.

## ONLINE
These are some of the websites where you can browse job ads or place your own ad:

- Production Base (www.productionbase.co.uk)
- Film-tv.co.uk (www.film-tv.co.uk)
- Grapevine Jobs (www.grapevinejobs.com)
- BBC Jobs (www.bbc.co.uk/jobs)
- Shooting People (http://shootingpeople.org)
- Media Guardian – Freelance CV service (www.mediaguardian.co.uk)
- StartinTV.com (www.startintv.com)
- Jobs in TV (www.jobsin.co.uk/tv)
- **IDEAS**FACTORY (www.channel4.com/ideasfactory)
- Workthing (www.workthing.com)
- Mandy's International Film and Television Production Directory (www.mandy.com)
- The Knowledge Online (www.theknowledgeonline.com)
- UK Screen (www.ukscreen.com)
- Regional Film & Video (www.4rfv.co.uk)
- Media Personnel (www.mediapersonnel.com)

## RECRUITMENT AGENCIES
New entrants can have a tough time persuading agencies to put them on their books, but it's worth a try:

- Production Switchboard – Tel: 01483 812 011
- Callbox Communications – www.callboxdiary.co.uk
- Switched On Production Consultancy – www.switchedonjobs.com

## Job ops for ethnic minorities
- The Cultural Diversity Network has an online database of black, Asian and other ethnic minority TV freelancers and staff. See www.channel4.com/diversity.
- Channel 4 sponsors four places, every two years, on the FT2 New Entrant Scheme. See www.ft2.org.uk.

## Job ops for people with disabilities
- Channel 4's website Four All is a tool for TV producers, broadcasters, casting agents and others who want to employ people with disabilities in their programmes – either on screen or behind the camera. See www.fourall.org.

- TV Futures – www.tvfutures.net
- Searchlight – www.search-light.com
- Cm Media – www.cm-media.co.uk
- Research Register (researchers) – Tel: 020 7700 7573
- Recruit Media (interactive) – www.recruitmedia.co.uk
- Mark Wilson (advertising) – www.mwrecruit.co.uk

### CVs
You're more likely to find work by sending out your CV to production companies and other potential employers, than answering a job ad that hundreds of other people have seen. Just take a look at the How They Got There table on page 88 – it's proof that most people working get their first break by contacting a company themselves.

Chances are your CV and letter will be opened, glanced at for about half a second and then put on a pile with dozens of others. It's still worth doing: when a production company has a sudden rush on, they'll go back to that pile of CVs, randomly pick a few up and call to see if the person is available to work – right now. That person could be you, so get your CV on those piles.

Nowadays it is also common to email your CV in. If you choose to do this, it may also be a good idea to make a follow-up phone call, as a busy potential employer may not have time to read your email as closely as you would like. Be aware that some employers may find unsolicited email approaches intrusive and hit the Delete button.

- The BBC gives advice and placements to people with disabilities, and has a Disability Programme Unit. See Get To Know Us at BBC Jobs (www.bbc.co.uk/jobs) or call the Diversity Unit (020 8576 1208).
- The Broadcasters and Creative Industries Network sponsors people with disabilities on the Production Base job search site (www.productionbase.co.uk), paying the annual fee for them. See the BCIDN page at the Employers' Forum on Disability (www.efd.org.uk/www/guests/bdn).

## Where to send CVs

- Think about programmes you love, find out who made them and send your CV to the company. The Internet is your best resource for this kind of research – just type the name of the show into Google (www.google.co.uk).
- Find out who's making what, and when – write to them asking about it. The only way to do this is to scour the trade press for small news briefs about who has been commissioned to make a programme or series.
- These are the reference books where you'll find names, addresses and numbers. Most of them are too expensive to buy, but you'll find them in your local library:
  *The PACT Directory of Independent Producers* (PACT)
  *The Knowledge* (CMP Information) – www.theknowledgeonline.com
  *Kays UK Production Manual* (Kay Media)
  *Kemp's Film and Television Handbook* (Reed) – www.kftv.com
  *BFI Film & Television Handbook* (BFI)
  *The Production Guide online –*
  www.broadcastnow.co.uk/ProductionGuide/
  *The Guardian Media Guide* (Atlantic Books)
  *Animation UK* (A-UK Publishing).

## DOS AND DONT'S

- Make your CV short – one side of A4 is enough.
- Do different CVs for different jobs – they need to be tailored for different jobs and companies.
- Go backwards – list your most recent jobs or education first.

## THE PERFECT CV

No, there isn't really any such thing as the perfect CV. But, to help you write a CV perfect for you, here's how skillsformedia suggest people lay them out:

Your name
Job title
Contact details
Personal profile
Key skills
Experience
Training
Personal details – age, interests, nationality, driving licence

*(Source: skillsformedia at www.skillsformedia.com)*

- There are no rules for laying out CVs, but in general it's best to highlight your experience before your education.
- Consider including a 'personal statement' or 'personal profile' at the top – a couple of lines summing up who you are and what skills and experience you have. For example: 'A Media Studies graduate, trained in camera work, with experience working as a Camera Assistant on professional and student projects.'
- It's nearly always better to send a sound, solid, confident CV than a supposedly zany, wacky one – unless you have a truly, mind-blowingly original idea or an astoundingly witty one.
- Get some advice: skillsformedia (Tel: 08080 300 900, www.skillsformedia.com) can give you help brushing up your CV.
- Channel 4's **IDEAS**FACTORY website also has a Plan Your CV section with useful advice and hints on creating the best CV possible. Go to www.ideasfactory.com/careers, then click on 'Knowledge', and you will find a CV-making tool.

### Covering letters
- Write a short, one-page covering letter. Make sure it's addressed to the right person – call beforehand to check exactly who you should send it to. Spell their name, and everything else in the CV and letter, correctly.
- The letter should have a simple format – say (briefly) why you're writing, who you are and what you've done, why you are writing to them in particular and that you'll hope they'll be in touch, or ask if you could meet.
- Make it clear in the letter what kind of work you're interested in. It drives people crazy when they get 'I'm-desperate-to-work-in-film, I'll-do-anything' letters.

### Calls
- Don't call to say you're sending your CV – they won't care.
- Do call to say you've sent it – make a couple of follow-up calls, one soon after you sent the CV, another later.

- On the phone, ask when would be a good time to call back – are they likely to be looking for staff in a few weeks or few months?
- Try to get chatting with the person on the other end, but also try to sound like you're only going to take 30 seconds of their time. Hearts sink when the person calling doesn't get to the point fast. So be quick.

## INTERVIEWS

- Feel confident. Make sure you're well prepared and have done your research. Then you can really *be* confident. If you've been asked in for an interview, they obviously think that, on paper at least, you're made of the right stuff.
- Research the company and its programmes, and go with opinions on them *and* ideas for new ones.
- First impressions matter. Smile, stand up straight. Ask someone in the industry what you should wear – you may not need to wear a suit, but you definitely still need to look as if you've made an effort.
- Shake hands with whoever's interviewing you. When someone asks you a question, look them in the eye when answering, but make sure you don't ignore anyone else present. If you feel nervous and twitchy, put your hands in your lap and keep them there.

# Future presenters

If you want to be a presenter, you'll need to make a showreel – a tape with you on – to send to casting agents and production companies. Here's Channel 4's advice:

- A showreel should be interesting, fun and certainly no more than 4–5 minutes in length
- Include several short scenes e.g. reporting from a special event or interviewing someone in a mock studio
- Remember the first minute is the most crucial to gain the viewer's attention
- Don't go to great expense – get family and friends to shoot it for you
- When you send your showreel, enclose a covering letter and ensure your name, age and contact details are clearly marked on the cassette. Never send the master copy
- Production companies may take some time to respond, especially if they are not currently casting.

*(Source: 4Careers at www.channel4.com/4careers)*

● You'll probably be asked if you have any questions, so prepare some beforehand. If you don't have any, it can look like you're not interested.

## Insider's tip

'If you're going to an interview with someone, find out who they are and what shows they've worked on – and mention them. TV people hate it more than anything when people haven't seen their work, so find out.'

*Graham Stuart, Executive Producer of* V Graham Norton
*and Director of* So Television

### WORK EXPERIENCE

You're much more likely to get a job, or get on a training scheme, if you've got some work experience. It shows you're committed to a career in TV and know something about the business, rather than just vaguely wanting to 'work in telly'. Not only that, work experience can lead to real work, or at least to making some important first contacts.

It also gives you a chance to see whether or not the TV world is for you. After one week of long hours, doing mind-numbing tasks and being treated like a nobody, you may decide it is not for you.
Finding work experience is not always easy. Production companies get hundreds of requests a year. To stand a chance, you need to treat looking

'Work experience can lead to real work, or at least to making some important first contacts. It also gives you a chance to see whether or not the TV world is for you.'

for work experience like job-hunting, which means researching which companies to approach and then sending well-written, easy-to-read CVs and covering letters.

That said, just because there is fierce competition to find work experience, it doesn't mean you should allow yourself to be exploited. Lots of people have work experience horror stories to tell, mostly about working for months on end, unpaid. It's very hard to say 'No' to a long stint of hands-on experience, even if it's unpaid. So set yourself a limit – decide how long you want, or can afford, to work unpaid and stick to it.

If you're doing well in a placement and they want you to stay on, tell them you'd love to, but you've got a time limit on unpaid work, and simply can't afford to do more without being paid. If they need you, they'll start paying.

### Finding work experience
- Use the Internet to check whether companies have work experience schemes. It will save you a lot of letter-writing time if you find out first that they *don't* take people on placements. Most companies only take people over 18.
- When writing letters or calling to ask for work experience, it may help to be specific – say what kind of experience you're after, in the production office or with the camera crew, rather than just that you want experience 'in TV'.
- In your letter, mention some of the programmes they make and say how much you like them. Call beforehand to find out who to send the letter to – ask who handles work experience requests.
- The BBC lets you search for placements online, in specific areas – production, engineering or finance, for example. See the Work Experience page at BBC Jobs (www.bbc.co.uk/jobs/workexperience_hub.shtml).
- Channel 4 offers a 'very limited number of placements'. Write to: HR Work Experience, Channel 4, 124 Horseferry Rd, London SW1P 2TX.
- Sky offers work placements in various areas. See www1.sky.com/jobs/workexperience.html.
- You can get work experience at film schools, working on student films. Some, like the National Film and Television School, have a work experience database. For more information, contact the NFTS (Tel: 01494 671 234; Website: www.nftsfilm-tv.ac.uk).
- Try the National Council for Work Experience (www.work-experience.org), which lets you search for placements online.

### NETWORKING
Networking... What does it mean? It sounds so formal, so phoney. Don't worry – networking just means talking to people, thinking of anyone you know who knows anything about TV, and then asking them for any advice or contacts they have. That's all. Make the most of the all contacts you have. Think about family, family friends, teachers, etc. and any connections

**Insider's tip**

'You have to be prepared to do anything, boring or not –
offer to make tea, photocopy, transcribe rushes, etc.
Don't come across as being arrogant to other members
of the team – there's nothing wrong with being confident
but over-confidence is over-bearing. Ask questions, no
matter how anal – don't pretend to know things, you'll
come unstuck in the end.'

*Sarah Roberts, Production Manager*

they may have with the industry. On reflection, it could turn out to be quite
an extensive network of useful contacts. Don't underestimate the value of
this kind of networking.

- Think of anyone you know who works in TV, or used to. If you don't know
  anyone, ask friends if they do. If you're feeling brave, you could contact
  people you have no connection with, but who might be able to give you
  some advice – look up some names in the TV reference books or find
  them in the credits. If you're feeling shy, contact people by email.
- If you get given someone's name and number, make sure you know who
  they are and what they do before you call. Introduce yourself, explain that
  'so and so' suggested you give them a call and ask if they've got time for
  a brief chat on the phone, or maybe even to meet for 15 minutes.
- If they say 'Yes', get as much information about a company, project, job
  or career as you can. You're basically asking them for advice on how to
  get ahead, the low-down on what steps you should be taking.
- If you're speaking to someone who might be in a position to offer you
  work, it's usually better to say you just want a 'chat', rather than saying
  you want a job. At the end, ask if it's OK to send your CV.
- Finally, if you still feel anxious about the thought of asking people for
  help, remember that on the whole, people *love* being asked for advice.
  It's very flattering, and a great chance to talk about themselves.

'If you started reading this book knowing you wanted to work in TV and, now you've got to the end, you still want to work in TV – go for it.'

## 7. GETTING ON
## ... is it for you?

Hopefully, this book will have given you a better idea of what it's like to work in TV and how to get into it. The big question now is... is it for you?

Do you have the talent and determination it takes to get training and your first break? Are you cut out for what can be a very insecure working life? Are you willing to work that hard? And do you love TV that much?

It may be that TV is almost, but not quite right for you. If so, perhaps you should think about a career in film or radio? Check out *Creative Careers: Film* and *Creative Careers: Radio*.

If you started reading this book knowing you wanted to work in TV and, now you've got to the end, you still want to work in TV – go for it.

There are interesting, exciting jobs out there, and a constant demand for new people with talent and energy to fill them. Working in TV can be tough – the hours can be long and there is not much stability – but if you're determined, hard-working and realistic, you'll have a great time.

## 8. FIND OUT MORE

Have a look over the On the Web boxes that have appeared throughout this book for a reminder of how much help and guidance you can find on the Internet.

## USEFUL ORGANISATIONS

### GENERAL
**Skillset and skillsformedia**
Prospect House
80–110 New Oxford Street
London WC1A 1HB
Tel: 020 7520 5757
Websites: www.skillset.org and www.skillsformedia.com

### PACT – Producers Alliance for Cinema and Television
45 Mortimer Street
London W1W 8HJ
Tel: 020 7331 6000
Website: www.pact.co.uk

**British Academy of Film and Television Arts (BAFTA)**
195 Piccadilly
London W1J 9LN
Tel: 020 7734 0022
Website: www.bafta.org

**Royal Television Society**
Holborn Hall
100 Gray's Inn Road
London WC1X 8AL
Tel: 020 7430 1000
Website: www.rts.org.uk

**OFCOM – Office of Communications**
*(A new body, replacing the Independent Television Commission, OFTEL, the Broadcasting Standards Commission and various radio authorities)*
Riverside House
Southwark
London SE1
Website: www.ofcom.org.uk

**BKSTS – The Moving Image Society**
Pinewood Studios
Pinewood Road
Iver Heath
Buckinghamshire SL0 0NH
Tel: 01753 656656
Website: www.bksts.com

**UNIONS, GUILDS AND TRADE ASSOCIATIONS**
**Advertising Producers' Association**
26 Noel Street
London W1F 8GT
Tel: 020 7434 2651
Website: www.a-p-a.net

**AMICUS – Amalgamated Engineering and Electrical Union**
Hayes Court
West Common Road
Hayes
Bromley
Kent BR2 7AU
Tel: 020 8462 7755
Website: www.aeeu.org.uk

**BECTU – Broadcasting, Entertainment, Cinematograph & Theatre Union**
373–377 Clapham Road
London SW9 9BT
Tel: 020 7346 0900
Website: www.bectu.org.uk

**Broadcast Journalism Training Council**
18 Miller's Close
Rippingale, near Bourne
Lincolnshire PE10 0TH
Tel: 01778 440025
Website: www.bjtc.org.uk

**Casting Directors Guild**
PO Box 34403
London W6 0YG
Tel: 020 8741 1951
Website: http://castingdirectorsguild.co.uk/

**Directors Guild of Great Britain**
Acorn House
314–320 Gray's Inn Road
London WC1X 8DP
Tel: 020 7278 4343
Website: www.dggb.co.uk

**Guild of British Camera Technicians**
GBCT, c/o Panavision UK
Metropolitan Centre
Bristol Road
Greenford
Middlesex UB6 8GD
Tel: 020 8813 1999
Website: www.gbct.org

**Guild of Location Managers**
20 Euston Centre
Regent's Place
London NW1 3JH
Tel: 020 7387 8787
Website: www.golm.org.uk

**Guild of TV Cameramen**
April Cottage
The Chalks
Chew Magna,
Bristol BS40 8SN
Website: www.gtc.org.uk

**Guild of Vision Mixers**
Website: www.guildofvisionmixers.co.uk

**Institute of Broadcast Sound**
27 Old Gloucester Street
London WC1N 3XX
Website: www.ibs.org.uk/

**Music Video Producers Association**
26 Noel Street
London W1F 8GT
Tel: 020 7434 2651
Website: www.mvpa.co.uk

**NUJ – National Union of Journalists**
Headland House
308 Gray's Inn Road
London WC1X 8DP
Tel: 020 7278 7916
Website: www.nuj.org.uk

**Production Guild of Great Britain**
Pinewood Studios
Pinewood Road
Iver Heath
Buckinghamshire SL0 0NH
Tel: 01753 651 767
Website: www.productionguild.com

**Production Managers Association**
Ealing Studios
Ealing Green
London W5 5EP
Tel: 020 8758 8699
Website: www.pma.org.uk

**Professional Lighting & Sound Association**
38 St Leonards Road
Eastbourne BN21 3UT
Tel: 01323 410335
Website: www.plasa.org

**Society of Television Lighting Directors**
Website: www.stld.org.uk

**TAC – Teledwyr Annibynnol Cymru**
Website: www.teledwyr.com

**Women in Film and Television**
6 Langley Street
London WC2H 9JA
Tel: 020 7240 4875
Website: www.wftv.org.uk

**Writers' Guild of Great Britain**
15 Britannia Street
London WC1X 9JN
Tel: 020 7833 0777
Website: www.writersguild.org.uk

## USEFUL BOOKS

*A Career Handbook for TV, Radio, Film, Video & Interactive Media*, by Shiona Llewellyn and Sue Walker (Skillset, A&C Black)

*The PACT Directory* (PACT)

*BFI Film & Television Handbook*, edited by Eddie Dyja (BFI)

*The Guardian Media Guide*, edited by Steve Peak (Atlantic Books)

**Trotman books on getting in to and working in the media**
*Creative Careers: Film*, by Milly Jenkins
*Creative Careers: Radio*, by Tania Shillam
*Q&A: Radio, Television & Film*, A Questions & Answers Careers Book
*Getting into the Media*, by Emma Caprez